Environmentally Friendly Syntheses Using Ionic Liquids

Sustainability: Contributions through Science and Technology

Series Editor: Michael C. Cann, Ph.D.
Professor of Chemistry and Co-Director of Environmental Science
University of Scranton, Pennsylvania

Preface to the Series

Sustainability is rapidly moving from the wings to center stage. Overconsumption of non-renewable and renewable resources, as well as the concomitant production of waste has brought the world to a crossroads. Green chemistry, along with other green sciences technologies, must play a leading role in bringing about a sustainable society. The **Sustainability: Contributions through Science and Technology** series focuses on the role science can play in developing technologies that lessen our environmental impact. This highly interdisciplinary series discusses significant and timely topics ranging from energy research to the implementation of sustainable technologies. Our intention is for scientists from a variety of disciplines to provide contributions that recognize how the development of green technologies affects the triple bottom line (society, economic, and environment). The series will be of interest to academics, researchers, professionals, business leaders, policy makers, and students, as well as individuals who want to know the basics of the science and technology of sustainability.

Michael C. Cann

Published Titles

Green Chemistry for Environmental Sustainability
Edited by Sanjay Kumar Sharma, Ackmez Mudhoo, 2010

Microwave Heating as a Tool for Sustainable Chemistry
Edited by Nicholas E. Leadbeater, 2010

Green Organic Chemistry in Lecture and Laboratory
Edited by Andrew P. Dicks, 2011

A Novel Green Treatment for Textiles:
Plasma Treatment as a Sustainable Technology
C. W. Kan, 2014

Environmentally Friendly Syntheses Using Ionic Liquids
Edited by Jairton Dupont, Toshiyuki Itoh, Pedro Lozano, Sanjay V. Malhotra, 2015

Sustainability: Contributions through Science and Technology

Series Editor: Michael C. Cann

Environmentally Friendly Syntheses Using Ionic Liquids

Edited by

Jairton Dupont
Institute of Chemistry, Universidade Federal do Rio Grande do Sul, Porto Alegre, Brazil

Toshiyuki Itoh
Graduate School of Engineering, Tottori University
Tottori, Japan

Pedro Lozano
Facultad de Química, Universidad de Murcia
Murcia, Spain

Sanjay V. Malhotra
Frederick National Laboratory for Cancer Research
Frederick, Maryland, USA

CRC Press
Taylor & Francis Group
Boca Raton London New York

CRC Press is an imprint of the
Taylor & Francis Group, an **informa** business

CRC Press
Taylor & Francis Group
6000 Broken Sound Parkway NW, Suite 300
Boca Raton, FL 33487-2742

First issued in paperback 2019

ISBN-13: 978-1-4665-7976-7 (hbk)
ISBN-13: 978-0-367-26254-9 (pbk)

Visit the Taylor & Francis Web site at
http://www.taylorandfrancis.com

and the CRC Press Web site at
http://www.crcpress.com

Contents

Preface

Increasing environmental consciousness within the scientific community has led chemists to search for environmentally friendly, nonpolluting media and processes for chemical synthesis as alternatives to conventional organic solvents. In the past two decades there have been numerous advances in organic synthetic methodologies that have made it feasible to avoid the use of toxic chemicals. Research in various chemical fields has produced substitutes for organic solvents and identified new materials such as ionic liquids (ILs). The unique properties of ILs, which can be tailored at the molecular level by an appropriate selection of its ionic units, have opened new avenues of processing options. Ionic liquids are now widely recognized as suitable for use in organic reactions and offer possibilities for improvement in the control of product distribution, enhanced reactivity, ease of product recovery, catalyst immobilization, and recycling.

This book reflects on the broad applications of ionic liquids and the latest developments in this field. The first chapter provides the historical perspective, including background and scientific observations that led to the development of the current field of ionic liquids. The focus of the second chapter is on organic syntheses in which ionic liquids have been used as reaction media. It provides examples of several chemical transformations that are commonly used in building complex chemical structures. This chapter also describes the role of ionic liquids as media for both stoichiometric and catalytic reactions. It explains the design of task-specific ionic liquids and provides information on the use of ionic liquids as activators of chemical reactions. Ionic liquids offer new possibilities for the application of solvent engineering to biocatalytic reactions. Criteria for developing biocatalytic processes using ionic liquids are described in Chapter 3. Several examples have been given to show how catalytic properties of enzymes are improved by using ILs as reaction media. ILs stabilize enzymes even under extremely harsh conditions, thereby bringing added value to the construction of a sustainable chemical industry. Chapter 4 details how ionic liquids combined with deep eutectic solvents provide valuable solutions for biotransformations. Given the lack of information about the toxicity of ionic liquids, their utility in health-related applications has been limited. Chapter 5 describes the ongoing effort exploring the role of ILs in the synthesis of pharmaceuticals and important building blocks, their use in drug delivery systems, and as active pharmaceutical ingredients. In the concluding chapter, the authors provide various examples, making a strong case for ILs as important liquid platforms for a range of catalytic processes.

Growing interest in the field of ionic liquids will define newer, as of yet unexplored areas of applications, expanding the potential utility of these environmentally friendly chemicals. It is our hope that the information presented in this book will prove to be a resource, motivating readers to explore this field further.

Sanjay V. Malhotra
Toshiyuki Itoh
Jairton Dupont
Pedro Lozano

About the Editors

Jairton Dupont earned his PhD at the Université Louis Pasteur of Strasbourg (France), and after a period as a postdoctoral fellow at the University of Oxford (UK) he became a professor of chemistry at the Institute of Chemistry, UFRGS (Brazil). He has been an invited professor at ULP (France), Nuremberg–Erlangen (Germany) and Universidad de Alcala de Henares and Rovira I Virgili (Spain). He is a member of the Brazilian Academy of Sciences. His various distinctions include the Humboldt Award, the Conrado Wessel Science Award, TWAS Award in Chemistry, and the Brazilian Gran Cruz. Dr. Dupont is currently associate editor of the *New Journal of Chemistry* (RSC). His research interests mainly center on ionic liquids with a special emphasis in catalysis, nanomaterials, and alternative energies. He has coauthored over 240 scientific publications and several patents and book chapters.

Toshiyuki Itoh graduated from Tokyo University of Education in 1976. After working as a chemistry teacher at Mie Prefectural High School, he returned to university and earned his PhD in 1986 from the University of Tokyo. He was appointed as an assistant professor at Okayama University in 1987 and promoted to associate professor in 1990. He worked with Professor Anthony G. M. Barrett as a postdoctoral fellow at Colorado State University from 1990 to 1991 and then moved to Tottori University as a full professor in 2001. Dr. Itoh is a recipient of the Society of Synthetic Organic Chemistry Japan Award (2010) and the Green and Sustainable Chemistry Award (2009). He is now serving as president of the Society of Fluorine Chemistry in Japan. His present major fields in chemistry include fluorine chemistry, ionic liquids chemistry, and the development of new and useful synthetic methodologies based on green chemistry.

Pedro Lozano earned his PhD in sciences (chemistry) from the University of Murcia, Spain, in 1988. From 1990 to 1991 he worked at the Centre de Bioingénierie Gilbert Durand, Toulouse, France, as a postdoctoral fellow under the supervision of Professor Didier Combes. In 1993, he returned to the Faculty of Chemistry at the University of Murcia as lecturer in biochemistry and molecular biology, and was promoted to full professor in 2004. Since 1996, he has been assistant dean of the faculty of chemistry and coordinator of the biochemistry degree at the University of Murcia. Dr. Lozano's research activity has always been related to enzyme technology under nonconventional media. His particular research interest focuses on the use of enzymes in neoteric solvents, i.e., ionic liquids and supercritical fluids. He is the author of more than 100 book chapters, papers, and journal articles that have appeared in such publications as *Green Chemistry, ChemSusChem, Energy & Environmental Sciences, Journal of the Supercritical Fluids,* and *Bioresources Technology.*

Sanjay V. Malhotra is the principal scientist and head of the Laboratory of Synthetic Chemistry at the Frederick National Laboratory for Cancer Research in Frederick, Maryland. He earned his PhD at Seton Hall University in 1995 and did postdoctoral research with Nobel Laureate Professor Herbert C. Brown at Purdue University. Subsequently, he became a university professor at the New Jersey Institute of Technology. In 2007 he moved to the National Cancer Institute at Frederick, Maryland, and established the Laboratory of Synthetic Chemistry. Dr. Malhotra's research interests include the design and synthesis of small molecules, and the development of new methodologies and materials for applications in drug discovery, environment, and energy.

Contributors

Juana M. Bernal
Facultad de Química
Universidad de Murcia
Murcia, Spain

Pablo Domínguez de María
Sustainable Momentum, SL
Canary Islands, Spain

Jairton Dupont
Institute of Chemistry
Universidade Federal do Rio
Grande do Sul
Porto Alegre, Brazil

Eduardo García-Verdugo
Universitat Jaume I
Campus del Riu Sec
Castellón, Spain

Toshiyuki Itoh
Graduate School of Engineering
Tottori University
Tottori, Japan

Pedro Lozano
Facultad de Química
Universidad de Murcia
Murcia, Spain

Santiago V. Luis
Universitat Jaume I
Campus del Riu Sec
Castellón, Spain

Sanjay V. Malhotra
Frederick National Laboratory
for Cancer Research
Frederick, Maryland

Brenno A. D. Neto
Instituto de Química
University of Brasília
Brasília, Brazil

Jackson D. Scholten
Institute of Chemistry
Universidade Federal do Rio
Grande do Sul
Porto Alegre, Brazil

Paulo A. Z. Suarez
Instituto de Química
University de Brasilia
Brasília, Brazil

Michel Vaultier
Université Bordeaux 1
CNRS-UMR
Talence, France

Hua Zhao
Chemistry Program
Savannah State University
Savannah, Georgia

chapter one

Introduction

Toshiyuki Itoh

Contents

The chemistry of ionic liquids (ILs) has an interesting history. Walden and coworkers reported that ammonium nitrate was present as a liquid at room temperature in 1914 [1], and the same results were reported by Sugden and Wilkins in 1929 [2]. However, no one paid any attention to these reports for half a century. In 1975, Osteryoung and his colleagues prepared halogenoaluminate(III) ionic liquids, such as ethylpyridinium bromotrichloroaluminate ([C_2Py][$AlBrCl_3$]), which was liquid at room temperature, and investigated their chemical and electrochemical properties [3]. The halogenoaluminate(III) ionic liquids are corrosive to many materials, and care must be taken in the selection of equipment: the equipment must be thoroughly cleaned and dried before use because it is easily decomposed by moisture and generates hydrogen chloride. These liquids were indeed difficult to handle under atmospheric conditions. Therefore, the early work reported by Osteryoung's group gained only a little attention, and it was 30 years before the brilliant future of ILs was recognized. A few chemists were impressed by the special nature of these salts, however, and attempted to use them as solvents for organic and inorganic reactions. Hussey first gave the term *ionic liquids* to these special molten salts in 1983 [4]; Wilkes and coworkers first demonstrated the Friedel–Crafts reaction using 1-butyl-3-methylimidazolium tetrachloroaluminate ([C_4mim][$AlCl_4$]) as both a solvent and a Lewis acid catalyst in 1986 [5]. However, we had to wait until 1992 to recognize the rich possibility of room temperature ionic liquids when Wilkes and Zaworotoko reported preparation of [C_2mim][BF_4] and [C_2mim][OAc], both of which are stable to moisture [6]. In 1996 Dupont and coworkers developed an important ionic liquid, 1-butyl-3-methylimidazolium hexafluorophosphate ([C_4mim][PF_6]), which shows a hydrophobic nature [7], while all room temperature molten salts reported previously had been hydrophilic and completely dissolved in water. These salts allowed us to handle them without special knowledge of the field, which strongly stimulated synthetic chemists to apply them as solvents for chemical reactions. In 1999, Welton wrote an important review of ILs in which he proposed using *ionic liquids* as the formal term of these interesting salts [8]; this term was quickly approved by chemists and became popular, while previous authors had been free to choose any name they wished for their systems, such as "liquid organic salts," "fused salts," and "room temperature molten salts."

Typical examples of ILs are shown in Figure 1.1. There have been reported numerous types that consist of a combination of various cations with anions [8, 9].

Design of the physical properties of ILs such as viscosity and conductivity is challenging, and no general rule has been found to date. However, for control of hydrophobicity of ILs, successful examples have been reported. It is well recognized that most ILs are hydrophilic liquids. For

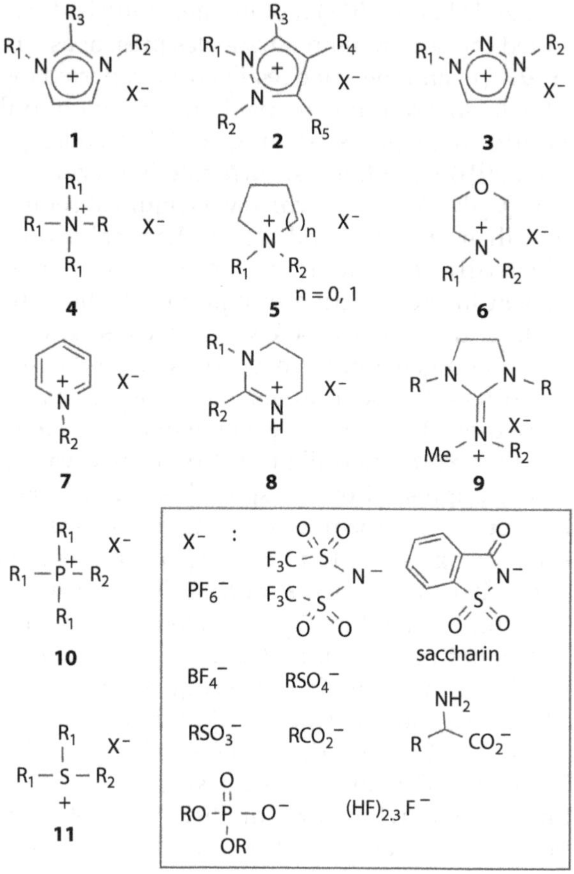

Figure 1.1 List of possible combinations of cations with anions that show liquids at RT.

example, those consisting of tetrafluoroborate, alkylsulfate, alkylsulfonate, carboxylate, or phosphate anions (see Figure 1.1) are hydrophilic and are completely dissolved in water. Dupont and coworkers first reported stable hydrophobic imidazolium salt, 1-butyl-3-methylimidazolium hexafluorophosphate ([C_4mim][PF_6]) in 1996 [10]. Since then, this salt has been used as a typical hydrophobic IL in many chemical reactions because its mixture with water forms a biphasic layer. Furthermore, this IL shows poor solubility in hexane or ether, which allows realization of an easy work-up process. However, [C_4mim][PF_6] was reported to be sensitive to the moisture at high temperature and produced hazardous hydrogen fluoride as it decomposed [11]. Therefore, bis(trifluoromethanesulfonyl)amide (NTf_2) salts are now recommended as the anion for preparing hydrophobic ILs.

Itoh and coworkers reported a successful example of a design of hydrophobic ILs using fluorine engineering [12]. They prepared three types of 1-butyl-3-methylimidazolium alkylsulfate: a 1:1 biphasic layer was obtained for 2,2,3,3,4,4,5,5-octafluoropentylsulfate ([C_4mim][C5F8]) due to its hydrophobic property (Figure 1.2, center), while n-pentylsulfate salt ([C_4mim][C5F0]) was completely dissolved in the water (Figure 1.2, left), and 4,4,5,5,5-pentafluoropentylsulfate salt ([C_4mim][C5F5]) showed a half hydrophobic property (Figure 1.2, right) [12].

Hagiwara and coworkers reported preparation of unique imidazolium salts with fluorohydrogenate [$(FH)_{2.3}F$]: these salts showed a very low melting point with high electric conductivity [13]. Fluorohydrogenate is a mixture of two anions of $(FH)_2F^-$ and $(FH)_3F^-$ = 7:3, and 1-ethyl-3-methylimidazolium fluorohydrogenate ([C_2mim][$(HF)_{2.3}F$]) shows a melting point at –65°C and conductivity at 100 mS cm^{-1} with a viscosity of 4.9 mPa s. Fluorohydrogenete ionic liquids are known as ILs that have the lowest melting points and viscosity with the highest electric conductivities to date.

On the other hand, the biological properties of ILs are dependent on both their cations and anions [14]. Generally, ILs with long alkyl substituents display antimicrobial activity and weak mutagenic properties toward plants [14]. However, many ILs reportedly have shown no significant toxicity for animals [14]. Interestingly, as expected, ILs formed with benzo[*d*]isothiazol-3-olate 1,1-dioxide anion (saccharin) (see Figure 1.1) are sweet [15].

The following features are recognized to be typical properties of ILs [8, 9]:

1. They are less volatile and flammable liquids. This means that we can avoid an explosion accident during chemical reactions if reactions are conducted in an IL.
2. ILs possess an extremely wide temperature range as liquid: they are generally thermally stable and are applicable to reactions under various temperature conditions.
3. Various kinds of organic and inorganic materials are dissolved in ILs.

4. They can be used as an aprotic highly polar solvent.
5. Many ILs display no serious toxicity for animals.

Solvent polarity is an important factor governing the chemical reaction because it has a strong impact on the reactivity of solute molecules. Since an ionic liquid is a salt, its polarity is considered to be very high. The log P value ($K_{o/w}$), which indicates the distribution efficiency of a liquid to n-octanol, has frequently been used as an index of solvent polarity [16]. However, it is obviously difficult to investigate log P values of a liquid with high-polarity solvents like an ionic liquid, and a method of estimating solvent polarity was thus developed using solvatochromism [17].

A betaine dye D (5′-(2,4,6-triphenylpyridin-1-ium-1-yl)-[1,1′:3′,1″-terphenyl]-2′-olate) in Figure 1.1 is a negative solvatochromatic compound (Figure 1.3), because the grand state energy of D is more decreased in a polar solvent due to solvation, while only a small stabilization occurs in the excited state (Figure 1.3) [17]. Therefore, a blue shift might be observed

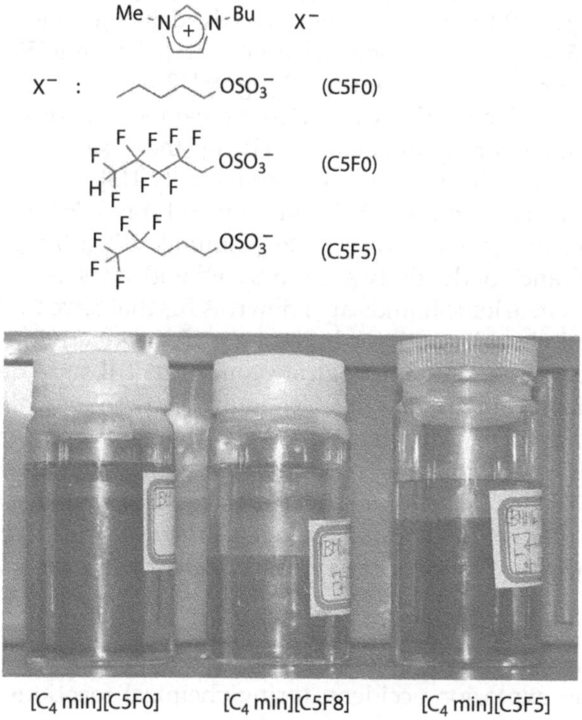

Figure 1.2 Hydrophobicity of three imidazolium salts. A 1:1 (v/v) mixture of IL with water was vigorously shaken and allowed to stand for 5 min.

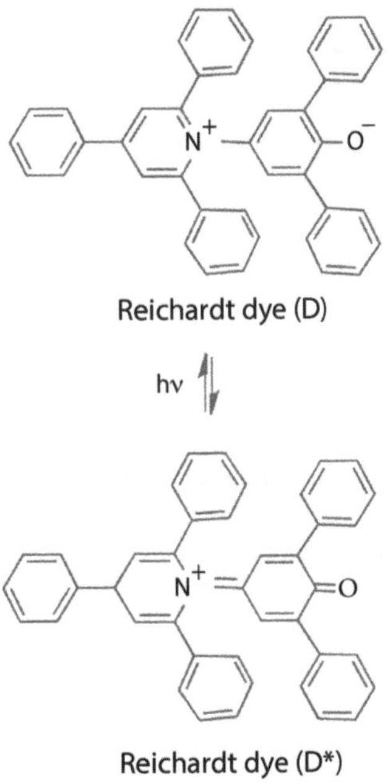

Figure 1.3 Reichardt dye and its structure in the excited state.

when D is dissolved in a highly polar solvent compared to a less polar solvent (Figure 1.4).

$$E_T{}^N = (E_T(\text{solvent}) - E_T(\text{TMS}))/(ET(\text{water}) - E_T(\text{TMS}))$$
$$= (E_T(\text{solvent}) - 30.7)/32.4 \quad (1.1)$$
$$E_T(\text{solvent})(\text{kcat mol}^{-1}) = 28{,}591/\gamma_{max}\ (\text{nm})$$

Reichardt and coworkers proposed the $E_T{}^N$ value as an index to estimate the polarity of a highly polar solvent using a betaine dye molecule (Reichardt dye): the $E_T{}^N$ value is calculated following Equation 1.1 [17].

Park and Kazlauskas investigated the polarity of various ILs using the $E_T{}^N$ values (Figure 1.5) [18]. As shown in Figure 1.5, the $E_T{}^N$ values of all ionic liquids tested were found in a range of 0.63 to 0.71. This level is similar to those of ethanol and slightly less than that of methanol. Although not as high as imagined by the word *salts,* it has been confirmed that the polarity of ILs is definitely very high: it is higher than acetonitrile or dimethyl sulfoxide (DMSO), which are well known to be the most polar aprotic solvents. Therefore, we now have numerous types of nonprotic polar liquids called ionic liquids at hand.

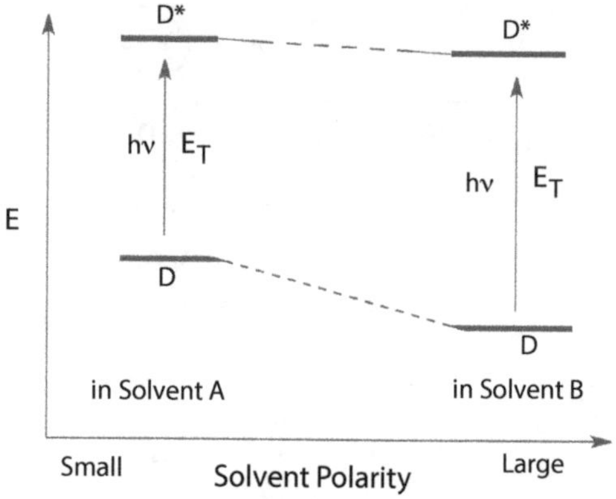

Figure 1.4 Solvatochromism of a betaine dye (Reichardt dye) and its use for estimating solvent polarity.

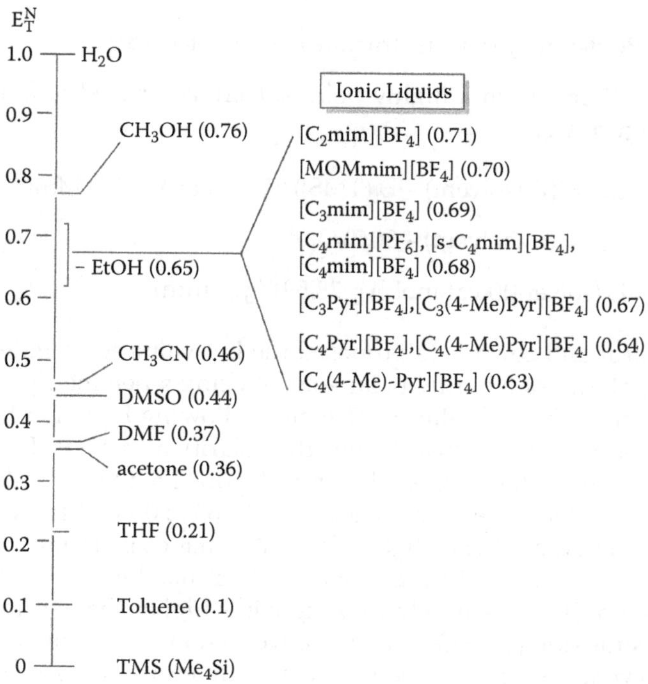

Figure 1.5 Polarity of various solvents using E_T^N scale. (From Park, S., and Kazlauskas, R. J., *J. Org. Chem.*, 66, 8395–8401, 2001.)

The solvent environment provided by ILs is quite unique compared to that of others available. As shown in Figure 1.6, there was a remarkably increasing number of publications in the field of ILs in 1999 and 2000. More than 20,000 papers have been published on ILs to date. But a turning point was reached in 1992, telling us that air-stable ILs such as [C_2mim][BF_4] and [C_4mim][PF_6] are the key ILs. A nice review by Welton [8] greatly stimulated many chemists and spotlighted studies in the ionic liquids field of chemistry (Figure 1.6). Although many reports have already appeared, the chemistry of ILs is still at an incredibly exciting stage in its development. In the following chapters, we will discuss the scope of ILs using typical examples of reactions in IL solvent systems from the standpoint of synthetic chemistry, in particular, green sustainable chemistry.

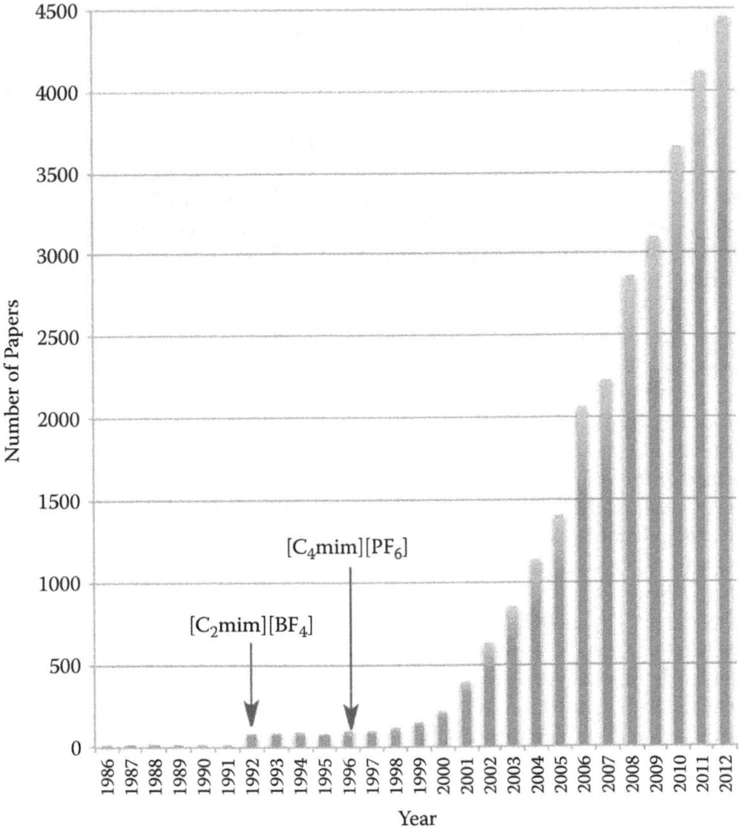

Figure 1.6 Publication history of ILs by a database provided from ISI (Web of Science). The darker gray color indicates papers in the field of organic and inorganic syntheses.

References

1. Walden, P. 1914. *Chem. Zentralbl.* 85: 1800–1801.
2. Sugden, S., and Wilkins, H. 1929. CLXVII. The parachor and chemical constitution. Part XII. Fused metals and salts. *J. Chem. Soc.* 1291–1298.
3. Chum, H. L., Koch, V. R., Miller, L. L., and Osteryoung, R. A. 1975. An electrochemical scrutiny of organometallic iron complexes and hexamethylbenzene in a room temperature molten salt. *J. Am. Chem. Soc.* 97: 3264–3265.
4. Hussey, C. L. 1983. Room-temperature molten salt systems. *Adv. Molten Salt Chem.* 5: 185–230; Hussey, C. L. 1988. Room temperature haloaluminate ionic liquids. Novel solvents for transition metal solution chemistry. *Pure Appl. Chem.* 60: 1763–1772.
5. Boon, B. J. A., Levisky, J. A., Pflug, J. L., and Wilkes, J. S. 1986. Friedel-Crafts reactions in ambient-temperature molten salts. *J. Org. Chem.* 51: 480–483.
6. Wilkes, J. S., and Zaworotoko, M. J. 1992. Air and water stable 1-ethyl-3-methylimidazolium based ionic liquids. *J. Chem. Soc. Chem. Commun.* 965–967.
7. Suarez, P. A. Z., Dullius, J. E. L., Einloft, S., De Souza, R. F., and Dupont, J. 1996. The use of new ionic liquids in two-phase catalytic hydrogenation reaction by rhodium complexes. *Polyhedron* 15: 1217–1219.
8. Welton, T. 1999. Room-temperature ionic liquids. Solvents for synthesis and catalysis. *Chem. Rev.* 99: 2071–2083.
9. Hallett, J. P., and Welton, T. 2011. Room-temperature ionic liquids: solvents for synthesis and catalysis. 2. *Chem. Rev.* 111: 3508–3576.
10. Suarez, P. A. Z., Dulius, J. E. L., Einloft, S., De Souza, R. F., and Dupont, J. 1996. The use of new ionic liquids in two-phase catalytic hydrogenation reaction by rhodium complexes. *Polyhedron* 15: 1727–1719.
11. Turner, M. B., Spear, S. K., Huddlestone, J. G., Holbrey, J. D., and Rogers, R. D. 2003. Ionic liquid salt-induced inactivation and unfolding of cellulase from *Ttichoderma reesei*. *Green Chem.* 5: 443–447.
12. Tsukada, Y., Iwamoto, K., Furutani, H., Matsushita, Y., Abe, Y., Matsumoto, K., Monda, K., Hayase, S., Kawatsura, M., and Itoh, T. 2006. Preparation of novel hydrophobic fluorine-substituted-alkyl sulfate ionic liquids and application as an efficient reaction medium for lipase-catalyzed reaction. *Tetrahedron Lett.* 47: 1801–1804.
13. Hagiwara, R., Matsumotoa, K., Nakamoria, Y., Tsuda, T., Ito, Y., Matsumoto, H., and Momota, K. 2003. Physicochemical properties of 1,3-dialkylimidazolium fluorohydrgogenate room-temperature molten salts. *J. Electrochem. Soc.* 150: D195.
14. Pham, T. P. T., Cho, C.-W., and Yun, Y.-S. 2010. Environmental fate and toxicity of ionic liquids: a review. *Water Res.* 44: 352–372.
15. Carter, E. B., Culver, S. L., Fox, P. A., Goode, R. D., Ntai, I., Tickell, M. D., Traylor, R. K., Hoffman, N. W., and Davis, J. H., Jr. 2004. Sweet success: ionic liquids derived from non-nutritive sweeteners. *Chem. Commun.* 630–631.
16. Leo, A., Hansche, C., and Elkins, D. 1971. Partition coefficients and their uses. *Chem. Rev.* 71: 525–615.
17. Reichardt, C. 1994. Solvatochromic dyes as solvent polarity indicators. *Chem. Rev.* 94: 2319–2358.
18. Park, S., and Kazlauskas, R. J. 2001. Improved preparation and use of room-temperature ionic liquids in lipase-catalyzed enantio- and regioselective acylations. *J. Org. Chem.* 66: 8395–8401.

chapter two

Organic synthesis using an ionic liquid as a reaction medium

Toshiyuki Itoh

Contents

Chemical reaction depends on three conditions: solvent, reaction temperature, and atmosphere. Among these conditions, the first thing we chemists should optimize is the solvent system. We have long been using two types of molecular solvent water and organic solvents, such as toluene, hexane, and ether. As mentioned in the introductory chapter, we are now able to choose a third type of liquid, ionic liquids (ILs), as chemical reaction media. ILs are completely identical with liquids of conventional molecular solvents, particularly in showing unique solubility vs. many organic and inorganic materials. This means that we can use ILs in various stages of chemistry. In this chapter, we show typical examples of synthetic reactions using an IL solvent system and discuss the scope of application of ILs as reaction media from the standpoint of sustainable organic synthesis.

2.1 *Milestone reaction in synthetic organic chemistry using an ionic liquid as a reaction medium*

The milestone reaction in the field of ionic liquids in synthetic organic chemistry might be the Mizorogi-Heck reaction reported in 1999 by Seddon and coworkers [1], though several early examples of organic reactions in ILs had already been reported previously [2] (Figure 2.1).

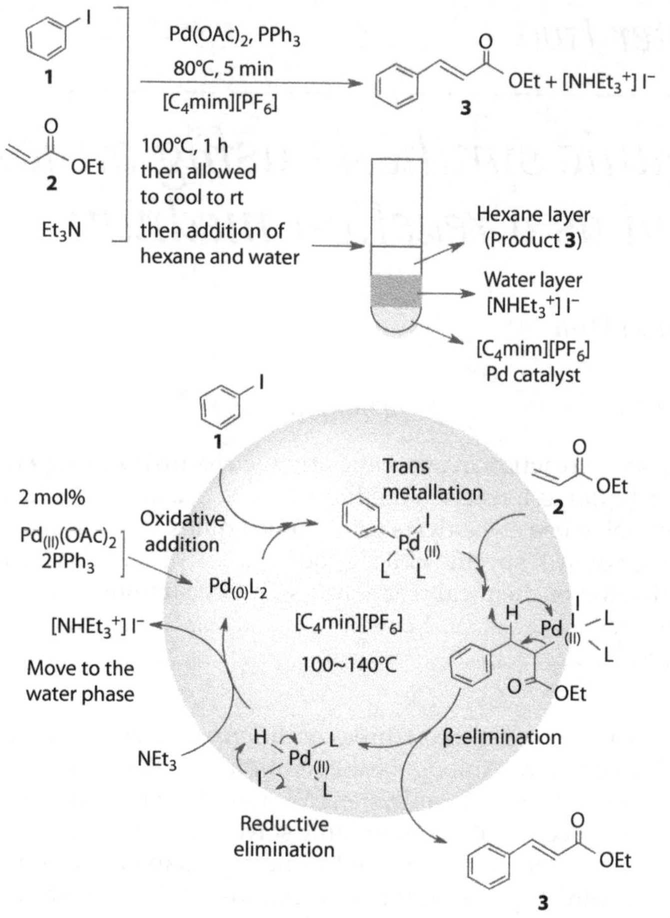

Figure **2.1** Milestone reaction using an IL as reaction medium (Mizorogi–Heck reaction).

A typical reaction was conducted as follows: a round-bottom flask equipped with a magnetic stirrer and reflux condenser was placed in [C_4mim][PF_6], palladium(II) acetate (2 mol% vs. substrate), and triphenylphosphine (2 eq. vs. Pd), and the mixture was heated to 80°C for 5 min with stirring to form the ionic liquid solution of the catalyst. A mixture of triethylamine (1.5 eq.), iodobenzene **1**, and ethyl acrylate **2** (1.25 eq.) was added, and the mixture was heated to 100°C for 1 h. The product was extracted from the reaction vessel by the addition of hexane to give *trans*-ethyl cinnamate **3** in excellent yield (>95%). In the reaction, triethylamine hydroiodide ([$NHEt_3^+$]I^-) was produced as a by-product (see Figure 2.1), which was removed easily by washing the reaction mixture with water, because triethylamine hydroiodide dissolved in water very well. On the other hand, the palladium catalyst remained in the ionic liquid layer after the work-up process. Therefore, after drying of the IL layer under reduced pressure, feeding a new set of substrates **1**, **2** and triethylamine caused a second reaction and the resulting product **3** was obtained without any reduction in chemical yield (Figure 2.1).

The authors thus succeeded in showing that the Pd catalyst and solvent could be recycled and reused six times without loss of activity of the catalyst. The most important feature of the reaction is realization of the immobilization of the transition metal catalyst in a solvent, and the reaction proceeded in the homogeneous state with high reaction efficiency. The recovery and reuse of catalysts have attracted a large interest in meeting the need for an environmentally benign reaction process. Homogeneous catalysis has an advantage in its ability to display full catalytic activity, but is accompanied by the drawback of laborious separation and recycling of catalyst. Immobilization of catalyst on a solid stationary support is a common method of realizing easy catalyst recovery. However, the resulting heterogeneous catalysts showed reduced catalytic activity and reaction efficiency under the heterogeneous reaction state was generally not sufficient. Achieving a recyclable use system of homogeneous transition metal catalysts has been a challenging project in organic synthesis. The IL solvent clearly provides such a reaction system. The work not only clearly indicates that an ionic liquid is just an alternative reaction medium of conventional organic solvents, but also provides a new possibility of organic syntheses.

2.2 *Recyclable use of catalysts using ionic liquids*

The IL solvent system allows us to use the homogeneous catalyst repeatedly. Sodeoka and coworkers reported excellent examples of enantioselective Pd-catalyzed reactions following this concept (Figure 2.2) [3]. They developed two types of reactions, enantioselective fluorination and a Michael reaction of β-ketoester derivatives **4** in the presence of a chiral palladium complex using an ionic liquid solvent system. In the case of asymmetric fluorination using *N*-fluoro-*N*-(phenylsulfonyl)benzenesulfonamide (NFSI), after completion of the reaction, the product was simply extracted with ether. The desired fluorinated product **5** was obtained in 93% yield with 92% ee, and recyclable use of the catalyst was also accomplished. Although a slight decrease of the reaction rate was observed in the 10th reaction cycle, this problem was solved by prolonging the reaction time, and the authors succeeded in obtaining **5** in 82% yield with 91% ee even after 11 repetitions of the process. Using the same catalyst, a Michael reaction of β-ketoester with methyl vinyl ketone has also been accomplished, and the desired adduct **6** was obtained in good yields.

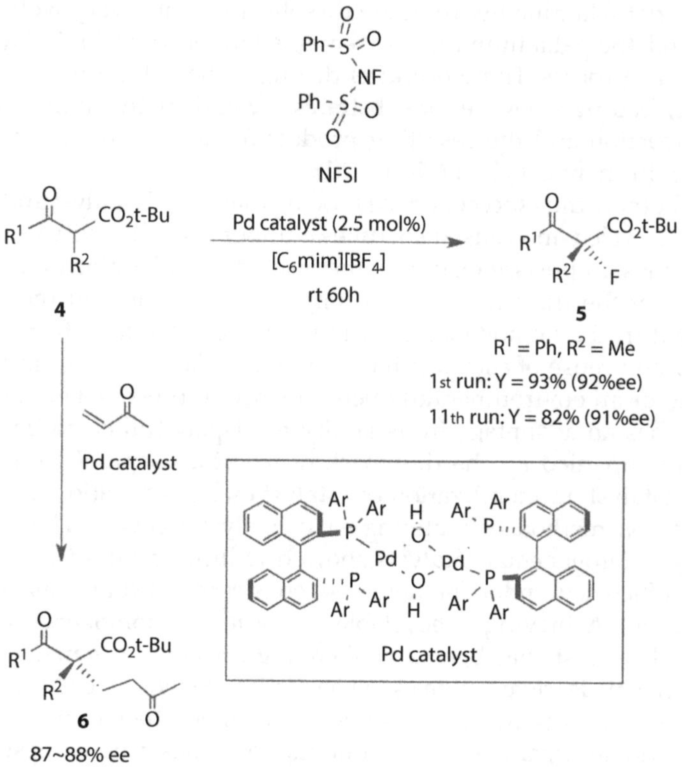

Figure **2.2** Pd-catalyzed enantioselective fluorination and Michael reaction in an IL.

Although the catalytic center of transition metal complexes is electrophilic, due to weak nucleophilicity, ionic liquids generally cause no reaction with the catalyst. An *N*-heterocyclic carbene (NHC) complex with transition metal sometimes forms in the presence of a weak base when using imidazolium type ionic liquids. Although the NHC complex generally contributes to the stabilization of zero valent transition metal, it sometimes inhibits the desired reaction. In the reaction system described in Figure 2.2, a neutral aqua Pd complex was used as a catalyst, and it has been shown numerous times that Pd catalyst works well without significant drop of the reactivity.

Itoh and coworkers also reported a good example of a recycling use system of transition metal catalyst in the IL reaction medium: the authors accomplished a sequence type of Nazarov cyclization and Michael reaction of pyrrole derivatives using iron(III) perchlorate, particularly, an alumina-supported one ($Fe(ClO_4)_3$•Al_2O_3) (Figure 2.3). The desired reaction of pyrrole 7 proceeded smoothly in the presence of 5 mol% of

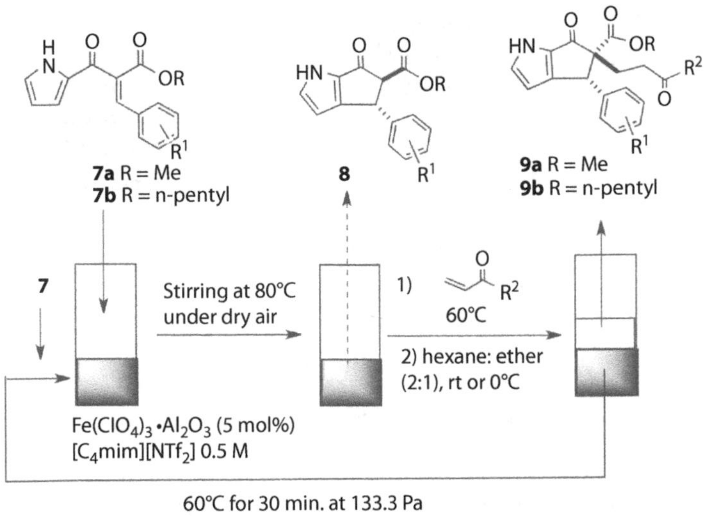

Figure 2.3 Recyclable use of Fe catalyst for one-pot sequential type Nazarov–Michael reaction using IL solvent system.

$Fe(ClO_4)_3$•Al_2O_3 in several IL solvents to afford the corresponding cyclized product **8**. Product **8** was obtained in excellent yield in every run five consecutive times and in acceptable yield nine times, although the reaction speed dropped with repetition of the reaction process, because a slight deactivation of iron catalyst took place due to moisture during the extraction process. They next attempted to use ILs as a solvent for the one-pot Nazarov–Michael reaction. The desired reaction indeed took place; however, the authors encountered unexpected difficulty: isolation of the final product, 4,5-dihydrocyclopenta[*b*]pyrrol-6(1*H*)-one derivative (**9a**, R = Me), from the reaction mixture was unsuccessful because **9a** was very soluble in [C_4mim][NTf_2]. Although this type of trouble sometimes happens using the IL solvent system, the authors solved it by switching the substrate from methyl ester to n-pentyl ester; they then succeeded in isolating the desired product, **9b** (R = n-pentyl), from the ionic liquid reaction mixture in acceptable yield (Figure 2.3) [4]. They next demonstrated the catalyst recyclable use a system employing n-pentyl ester in [C_4mim][NTf_2] as solvent.

Asymmetric reduction of ketones using a transition metal catalyst is now widely recognized as an important method for preparing optically active alcohols. Since a chiral transition metal catalyst is expensive, recyclable use of a catalyst using ILs provides a very attractive means in this field. Lin and coworkers reported a good example where they accomplished asymmetric reduction of aromatic ketones **10** with chiral ruthenium complex (*R,R,R*)-**12**: the resulting chiral alcohols **11** were obtained with excellent enantiomeric excess (Figure 2.4). Hydrophobic ILs such as

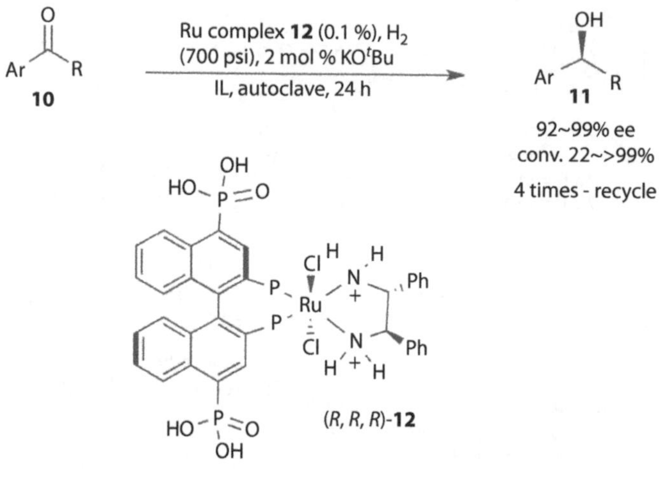

Figure 2.4 Asymmetric reduction of ketone with chiral Ru complex and recyclable use of the catalyst in the IL solvent system.

[C_4mim][PF_6] or [C_4dmim][NTf_2] worked as a good solvent, and a mixed solvent with 2-propanol with these ILs attained high enantioselectivity [5].

Wasserscheid and his colleagues reported an asymmetric reduction of β-ketoester with different ruthenium complex with chiral binaphthyl ligand (Figure 2.5) [6]. In this system, they revealed that it was essential to use a mixed solvent of an IL with methanol for obtaining high enantioselective reaction. They supposed the reason why methanol was necessary for the reaction, as illustrated in Figure 2.5: the reaction rate of direct reduction of the carbonyl group from **11** to **14** proceeded slowly; the key step of the enantioselective reaction might be considered an enantioselective conversion of intermediate **16**, and addition of methanol greatly accelerated the path of **15** to convert **16**. Therefore, appropriate choice of cationic group of the ILs was important to obtain high enantioselectivity, and [bis(hydroxyethyl) dimethylammonium][NTf_2] gave the best result, while only poor enantioselectivity was attained for conventional [C_4mim][NTf_2] [6].

Since nitro group is easily converted into several other functional groups, development of chiral nitro alcohols is an important topic in the synthetic organic chemistry. It is well recognized that the Henry reaction provides nitro alcohols, and it requires highly polar solvents. Since ILs are highly polar solvent, they worked as good reaction media for the Henry reaction. Among developed catalysis for the Henry reaction, a copper complex-mediated catalyst might be one of the most attractive because it is inexpensive and less toxic in nature, but recyclable use of copper catalyst remains to be realized. Kahn and Park solved this problem using an IL solvent: an asymmetric Henry reaction of aldehyde **17** took place using 10 mol% of $Cu(OAc)_2•H_2O$ in the presence of ligand **19** in a mixed solvent

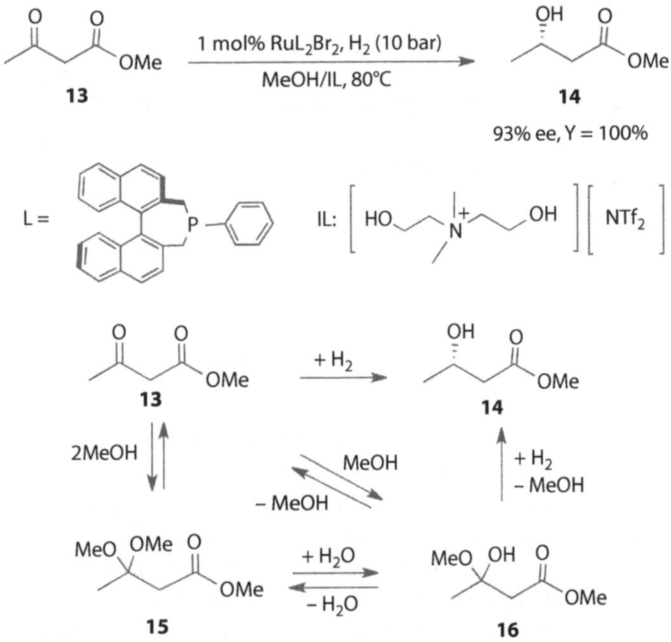

Figure 2.5 Asymmetric reduction of β-ketoester with chiral Ru complex using a mixed solvent of IL with methanol.

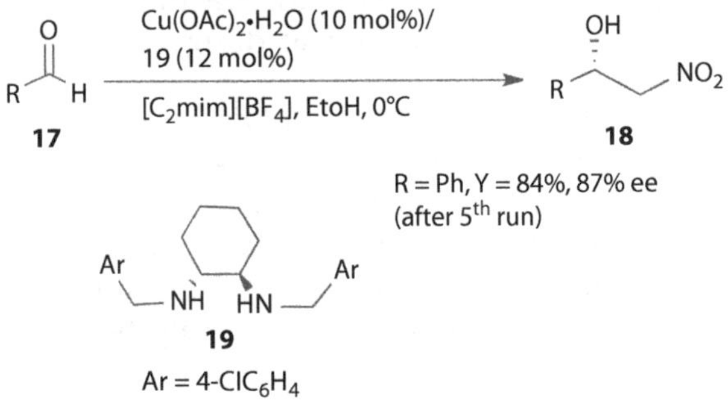

Figure 2.6 Chiral copper complex-mediated asymmetric Henry reaction using IL solvent system.

system of [C_2mim][BF_4] with ethanol and gave product **18** with high enantiomeric excess after five repetitions of the catalyst (Figure 2.6) [7].

The ionic liquid solvent system is applicable to organocatalyst-mediated reactions. Barbas and coworkers demonstrated an *L*-proline-mediated asymmetric Mannich reaction (Figure 2.7). Cyclohexanone **20** reacted with (Z)-ethyl 2-((4-methoxybenzyl)imino)acetate **21** in the presence of 20 mol% of *L*-proline, and the desired Mannich product **22** was

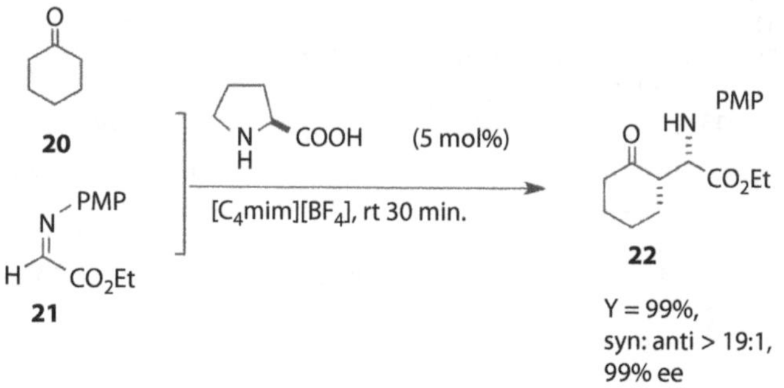

Figure 2.7 L-Proline-mediated asymmetric Mannich reaction in an IL solvent system.

obtained in 99% yield with >99% ee. Since L-proline is very soluble in $[C_4mim][BF_4]$, efficient recyclable use of the catalyst was accomplished [8].

Gruttadauria and coworkers reported that IL made it possible to realize recyclable use of the organocatalyst system (Figure 2.8) [9]. They demonstrated an L-proline-mediated asymmetric aldol reaction of **23**. Using this catalyst system, the authors accomplished the reaction in moderate enantioselectivity even after four recyclable uses of the catalyst. Addition of IL such as $[C_4mim][BF_4]$ was essential to achieve recyclable use of L-proline, and they revealed that it was unsuccessful to use L-proline repeatedly if the reaction was carried out only in the presence of L-proline. Although enantioselectivity of the reaction should be improved, the methodology provides a good hint to design an efficient recyclable organocatalysis system.

Numerous examples of recyclable use of a homogeneous catalyst have thus been shown using ionic liquid solvent systems. It should be emphasized that the reaction rate of a homogeneous catalyst system is generally superior to that of a solid catalyst reaction system. However, a problem has sometimes been observed that reaching the catalyst took place during the extraction process for homogeneous catalysts. Such a problem is solved by modification of the catalyst through introduction of an imidazolium salt moiety, which is generally called an IL tag.

A typical recent example is shown in Figure 2.9. Shimakoshi and coworkers demonstrated an interesting photoinduced aerobic oxidation of naphthalene-1,5-diol (**26**) to 5-hydroxynaphthalene-1,4-dione (**27**) in the presence of a catalytic amount of porphycen in an IL solvent system. However, since porphycen was very soluble in an extraction solvent such as ether, significant reaching into the extract took place during the work-up process. They solved this difficulty by modifying porphycen through the introduction of the imidazolium salt moiety (Figure 2.9) [10]. With this modification, they succeeded in demonstrating the recyclable use of the catalyst.

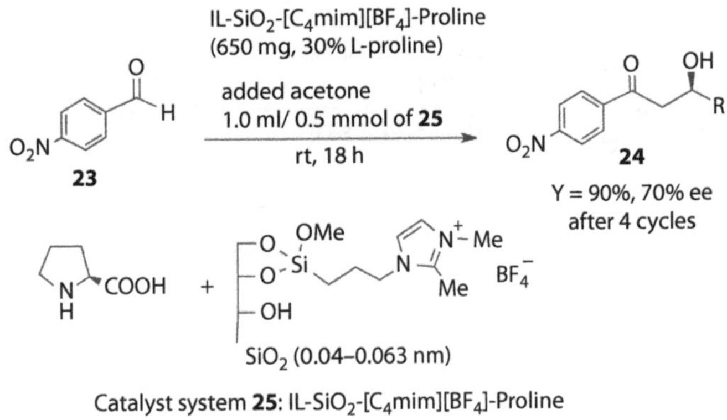

Figure 2.8 Asymmetric aldol condensation using an organocatalyst system.

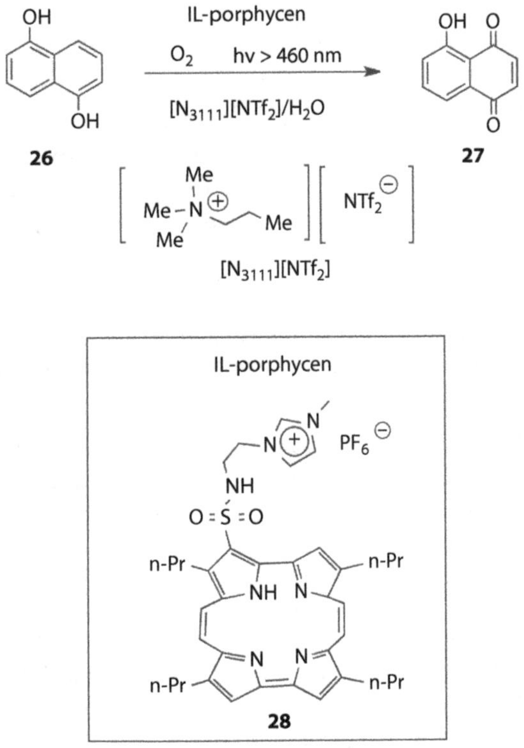

Figure 2.9 Photoinduced oxidation of naphthalene-1,5-diol using IL-porphycen.

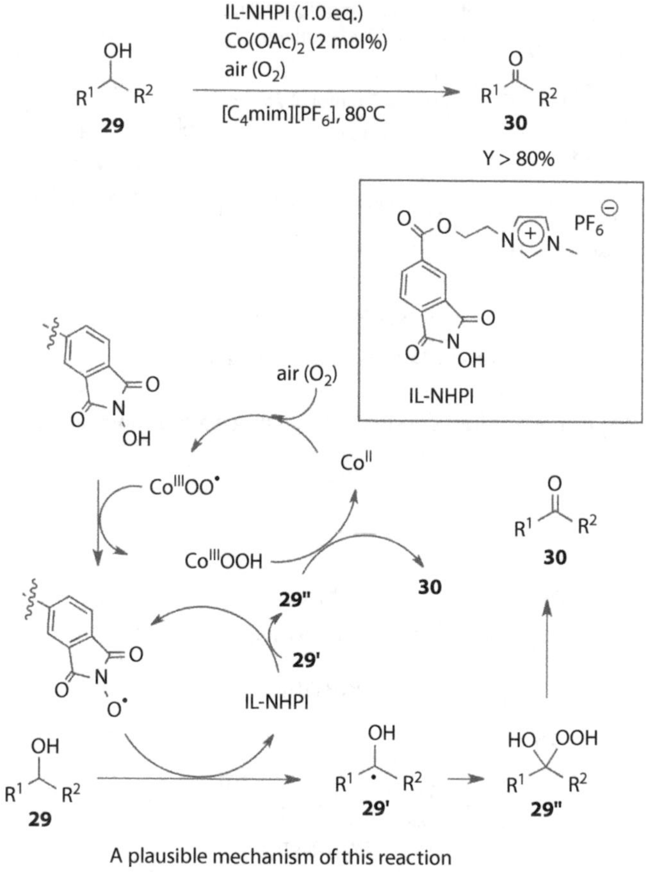

Figure 2.10 Recyclable use of IL-supported mediator IL-NHPI for $Co(OAc)_2$-catalyzed aerobic oxidation of secondary alcohols and its plausible reaction mechanism.

Figure 2.10 shows another example using the IL tag for organic synthesis in an IL solvent system. Kitazume and Koguchi prepared *N*-phthalimide-substituted imidazolium salt (IL-NHPI) and used it as a mediator of $Co(OAc)_2$-catalyzed aerobic oxidation of secondary alcohol **29** in [C_4mim][PF_6] solvent. Use of IL-NHPI significantly enhanced the reaction efficiency of $Co(OAc)_2$-catalyzed aerobic oxidation of various secondary alcohols in an IL solvent, and desired ketone **30** was obtained in excellent yield [11]. In the reaction, the IL solvent stabilized not only the cobalt catalyst, but also its mediator (NHPI). They therefore demonstrated recyclable use of the catalyst system five times after extraction of the product in a [C_4mim][PF_6] solvent system.

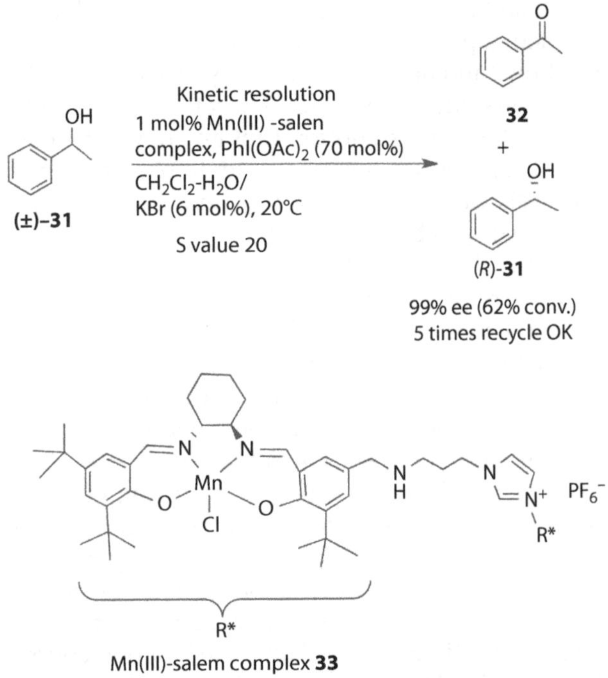

Figure 2.11 Mn-salen complex-mediated asymmetric oxidation of 1-phenylethanol (31).

Asymmetric oxidation of secondary alcohol using a chiral Mn(III)-salen complex has been reported by Li and coworkers (Figure 2.11) [12]. Racemic 1-phenyl ethanol ((±)-**31**) was treated with iodobenzen diacetate ($PhI(OAc)_2$) in the presence of 1 mol% of chiral Mn-salen complex **33**, and (*S*)-1-phenylethanol was selectively oxidized to afford ketone **32**. Since the reaction rate between the enantiomers was so large that kinetic resolution of (±)-**31** occurred during the reaction, they thus obtained alcohol (*R*)-**31** with 99% ee at 62% conversion (Figure 2.11). In this reaction, the imidazolium salt tag system was effectively used for recycling the catalyst. They demonstrated the reaction in a mixed solvent of dichloromethane with water, and then extracted ketone **32** was produced and (*R*)-1-phenylethanol (**31**) unreacted with hexane. Since the catalyst remained in the water layer due to the highly hydrophilic nature of the imidazolium group, they succeeded in demonstrating recyclable use of the catalyst five times without any loss of enantioselectivity.

Ionic liquids are now recognized as suitable for use in organic reactions and offer ease of product recovery, catalyst immobilization, and recycling. Recently, enzymatic reactions have become popular for preparing chiral compounds in industry. It has been well recognized that an enzymatic reaction proceeds in a buffer aqueous solution under appropriate pH conditions, and an enzyme quickly loses its activity in a highly concentrated aqueous salt solution [13]. Therefore, from the standpoint of biology, it seems to be a foolish notion that an enzymatic reaction occurs in a salt medium. Lipases are the most widely used enzymes applicable for organic synthesis. Lipase-catalyzed reactions in an ionic liquid solvent system have now been established [14, 15].

Is it possible to realize recyclable use of enzymes in an ionic liquid solvent system? For the recyclable use of lipases, immobilization of lipases on an appropriate solid support is well established; the enzyme can be used several months using the flow reaction system in some organic solvents [12]. However, the lipase reactivity quickly disappears when the solid-immobilized lipase is kept in an organic solvent without substrate esters or alcohols; furthermore, the reaction rate is generally insufficient. Itoh and coworkers succeeded in showing that recycling of the enzyme was indeed possible in an IL solvent system, and they accomplished recyclable use of lipase using ILs as solvent (Figure 2.12) [16]. Typically, the reaction was carried out as follows: to a mixture of lipase in the ionic liquid were added this racemic alcohol ((±)-**34**) and vinyl acetate as the acyl donor. The resulting mixture was stirred at 35°C and the reaction course was monitored and ether was added to the reaction mixture to form a biphasic layer. The product acetate (*R*)-**35** and unreacted alcohol (*S*)-**34** were extracted with ether quantitatively. Although the acylation rate was strongly dependent on the anionic part of the solvents, perfect enantioselective reaction was accomplished in the IL solvent system, and the enzyme remained in the ionic liquid phase as expected.

In the lipase-catalyzed transesterification reaction, vinyl acetate is generally used as the acyl donor because acetaldehyde is produced by the lipase-catalyzed transesterification, and this causes no reverse reaction. However, it is known that acetaldehyde acts as an inhibitor of enzymes because it forms a Schiff base with amino residue in the enzyme; since acetaldehyde easily escapes from the reaction mixture due to its very volatile nature, it therefore causes no inhibitory action on the lipase. Itoh and coworkers found that the reaction rate gradually dropped with repetition of the reaction process when using [C_4mim][PF_6] or [C_4mim][BF_4] as a solvent [16]; this drop in reactivity was caused by the inhibitory action of acetaldehyde oligomer, which had accumulated in the [C_4mim][PF_6] solvent system. This oligomer formation was assumed to be caused by water that was trapped by the 2-position of the imidazolium cation during the work-up process, because high acidity of the 2-position of the imidazolium cation was established [17]. This problem was effectively solved simply by the design of ILs (see the middle scheme in Figure 2.12): using 1-butyl-2,3-dimethylimidazolium ([C_4mim]), which lacked hydrogen at

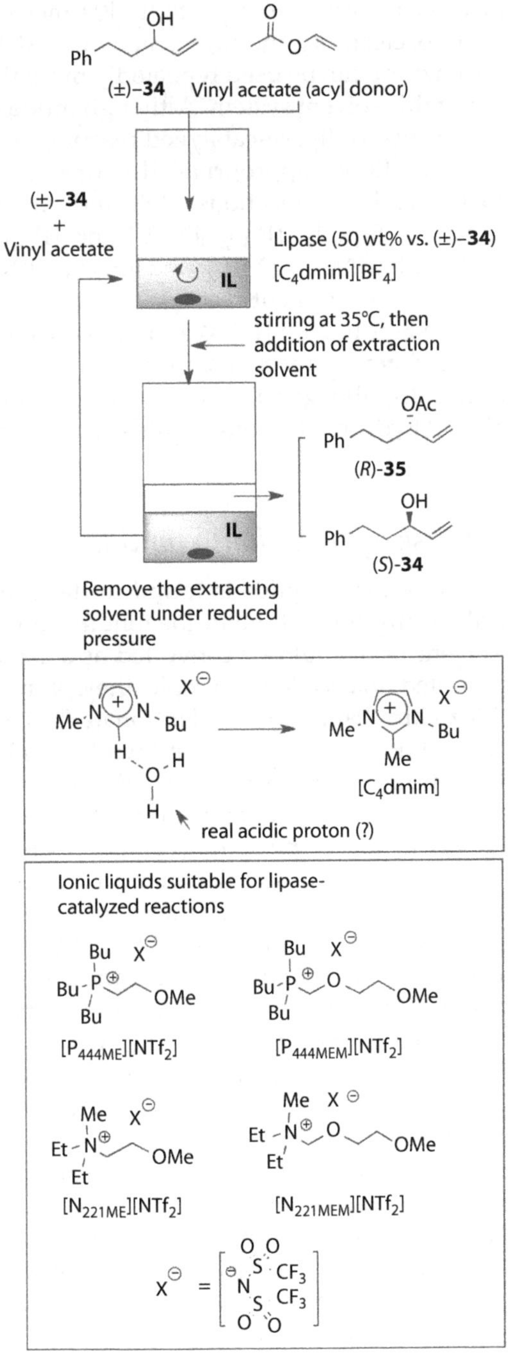

Figure 2.12 Lipase recyclable use system using an IL as reaction medium.

the 2-position, the enzyme was used repeatedly 10 times while still maintaining perfect enantioselectivity and high reactivity [18]. It has now been confirmed that the enzyme can be used repeatedly more than 20 times for several months using this solvent system. Although imidazolium salts are generally used as solvents for lipase-catalyzed reaction, the results clearly tell us that we should choose appropriate ILs for enzymatic reactions like transition metal-catalyzed reactions. Itoh and coworkers reported that phosphonium ionic liquids ([P_{444ME}][NTf_2] and [P_{444MEM}][NTf_2]) and ammonium ionic liquids ([N_{221ME}][NTf_2] and [N_{221MEM}][NTf_2]) that have alkylether moieties become excellent reaction media for lipase-catalyzed transesterification (see Figure 2.12) [19]. We can now use various types of ILs as solvents for biochemical reactions: imidazolium salts, ammonium salts, pyrrolidinium salts, alkylguanidinium salts, pyridinium salts, and phosphonium salt and hydrophobic ionic liquids are generally suitable for these reactions [19].

2.3 *Design of task-specific ionic liquids*

An important feature of ionic liquids is found in the versatility of their design. If chemical modification of the imidazolium frame is carried out, the ionic liquid solvent itself could be converted as a catalyst. Davis and coworkers demonstrated an excellent example: they prepared 3-methyl-1-(4-sulfobutyl)imidazolium trifluoromethylsulfonate ([$C_4(SO_3)$imim][OTf]), which involved a sulfonic acid side chain and worked as both a Brønsted acid catalyst and a solvent for esterification of carboxylic acid with alcohol; ester **35** was obtained in good yield, and after extraction of the product ester, the ionic liquid was used repeatedly (Figure 2.13, Equation 2.1) [20].

Preparation of Lewis acid IL is easier than that of Brønsted acid IL. As mentioned in the introduction, Wilkes and coworkers demonstrated the Friedel–Crafts type using [C_4mim][$AlCl_4$] as both solvent and acylation catalyst. However, this IL was difficult to handle under atmospheric conditions. Seddon and coworkers prepared a mixed type IL ([C_2mim][$AlCl_3I$]) simply by mixing imidazolium iodide with $AlCl_3$ and accomplished the desired acylation of toluene: reaction of toluene with acetyl chloride in the solvent system gave **21** in excellent yield with perfect regioselectivity (Figure 2.13, Equation 2.2) [21].

It is sometimes reported that the leak of the catalyst causes a serious problem. For example, selenium oxide is well known as an efficient catalyst of urethane synthesis through oxidative carbonylation of amines with carbon monoxide. However, a significant amount of toxic selenium oxide moves into the product. Kim and coworkers solved this difficulty using ionic liquid engineering [22]. They prepared imidazolium methylselenonate and used it as a catalyst of oxidative carbonylation of amines (Figure 2.13, Equation 2.3): significant reduction of the amount of selenium

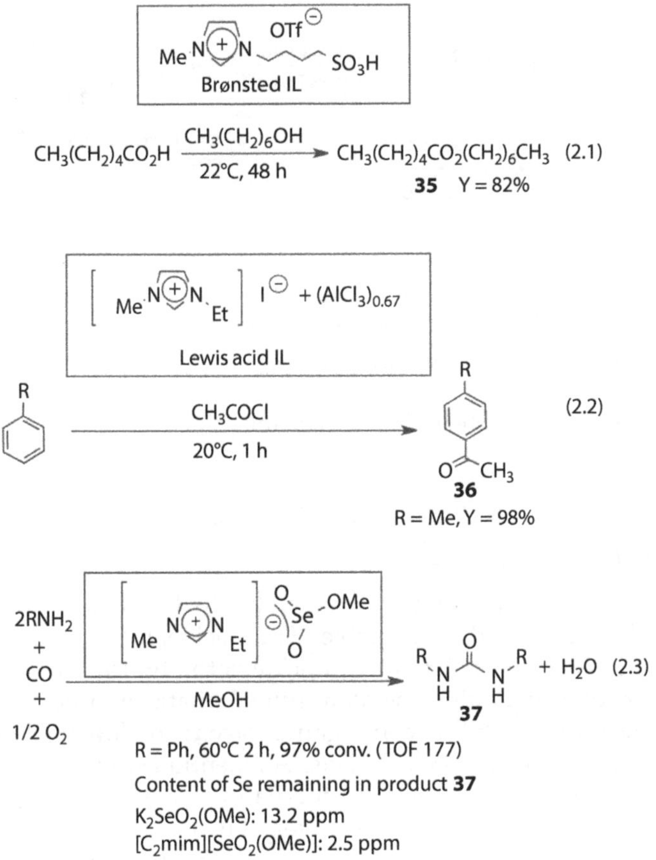

Figure 2.13 Design of ILs as both catalyst and solvent.

contamination in product **37** was accomplished. Although a high level of selenium contamination (13.2 ppm) was detected when the reaction was carried out using potassium dimethylselenate as catalyst, the amount of selenium contamination was reduced drastically to 2.5 ppm for [C_2mim][SeO_2(OMe)]-catalyzed reaction.

Luo and coworkers prepared chiral imidazolium salt ILs **39** and **40**, which were derived from *L*-proline, and used them as a phase transfer catalyst of asymmetric Michael reaction of 1-nitro-2-phenylethen with cyclohexanone: the corresponding 1,4-adduct **38** was obtained in excellent yield with 99% ee using 15 mol% of imidazolium salt **39** (Figure 2.14). The authors also demonstrated the reaction using isoquinolinium salt **40** in a [C_4mim][BF_4] solvent system as catalyst and demonstrated recyclable use of the catalyst [23].

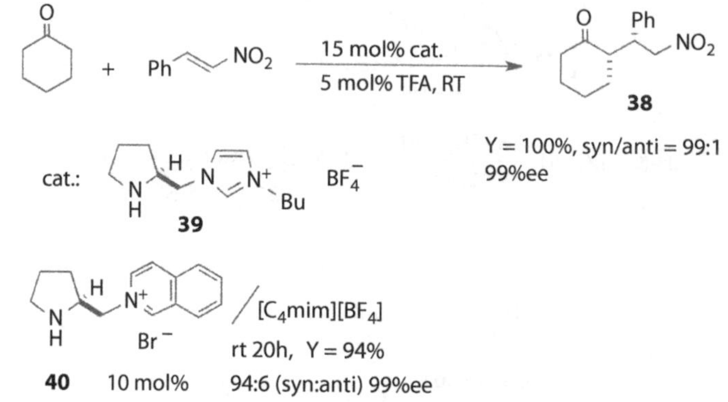

***Figure* 2.14** Asymmetric Michael reaction using chiral ILs.

2.4 *Activation of chemical reactions using ionic liquids*

The solvent environment that is provided by the ILs is quite unique compared to that of any other available. The nucleophilic displacement of various sulfonates and halides by fluoride is the typical method for the introduction of a single fluorine atom into aliphatic organic compounds. However, its limited solubility in organic solvent and low nucleophilicity require generally vigorous conditions. As mentioned, it has been established that ILs can act as powerful media in catalytic organic reactions not only for facilitation of catalyst recovery, but also for the acceleration of reaction rate and improvement of selectivity. Chi and coworkers reported a typical example of activation of SN_2 reaction of methanesulfonate **41** with potassium fluoride: facile nucleophilic fluorination of **41** to the corresponding fluoroalkane **42** was accomplished using potassium fluoride (KF) in an ionic liquid $[C_4mim][BF_4]$ [24] (Figure 2.15). They further found that addition of a small amount of water significantly improved reaction efficiency and that a mixed solvent of $[C_4mim][BF_4]$, acetonitrile (CH_3CN), and water (160:320:9) was extremely effective for the reaction affording **42** in excellent yield [24]. Inspired by the results, they conducted a detailed study on the origin of the activation of fluoride anion and found that addition of a small amount of *tert*-butylalcohol (*t*-BuOH) or *tert*-amyl alcohol was very effective in solvating cesium fluoride in an IL solvent and also resulted in the desired SN_2 type fluorination of methanesulfonate (see Figure 2.15). They successfully applied their reaction to the preparation of ^{18}F-subsituted medicinal compounds [25]. Chi's group prepared polymer-supported imidazolium salt **43** and accomplished activation of metal fluoride for a SN_2 type reaction with alkylmethanesulfonate [26].

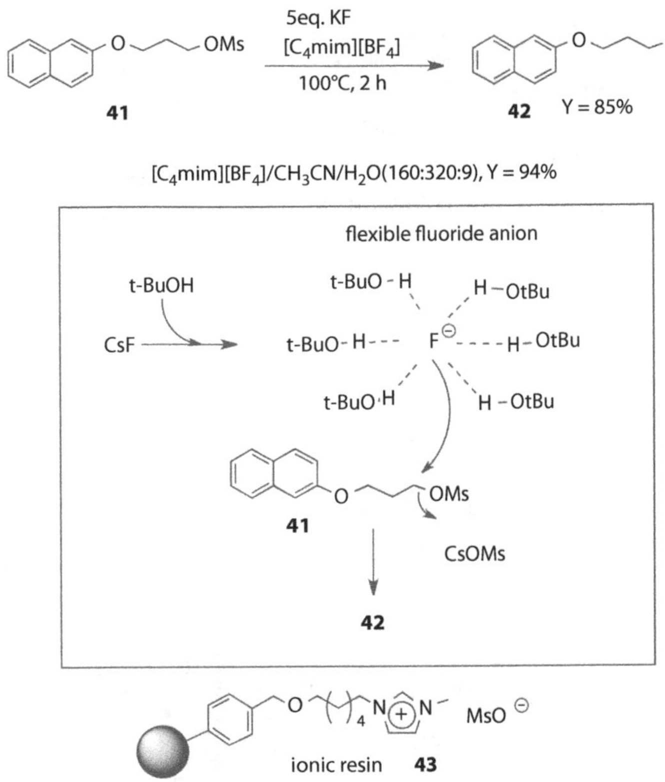

Figure 2.15 IL-mediated activation of SN_2 substitution of methanesulfonate with fluoride anion.

Ohara and coworkers reported that [2 + 3]-type cycloaddition of styrene derivatives with 1,4-benzoquinone took place in an IL solvent (Figure 2.16) [27]. This reaction also occurred in acetonitrile (CH_3CN), but there was a large difference in the reaction rate. The reaction in $[C_4mim][PF_6]$ proceeded very rapidly (ca. 40-fold) compared to that in CH_3CN under the same concentration [27]. It was revealed that the reaction started from a single electron oxidation of styrene derivative **44** by the ferric ion, and ferrous ion then generated reduced 1,4-benzoquinone **45**. The resulting cation radical **44*** reacted with benzoquinone radical **45*** to form [2 + 3]-adduct A, which immediately isomerized to give B, and then the final product **46** was produced [27]. The plausible mechanism of the reaction was supported by two facts: positive solvatochromism was observed during the reaction, and the reaction was inhibited by addition of a radical scavenger. Since ionic liquids are highly polar liquid, electron charge-separated intermediates like **44*** and **45*** might be produced more easily

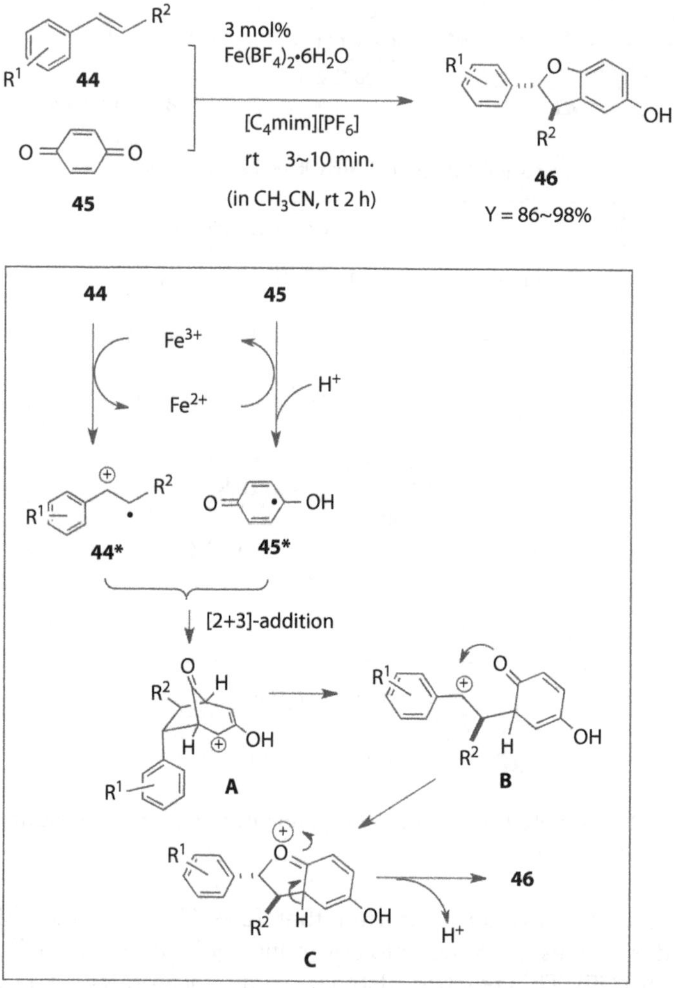

Figure 2.16 Iron(III)-catalyzed [2 + 3]-cycloaddition of styrene derivatives with 1,4-benzoquinone in an IL solvent and its plausible mechanism.

than CH_3CN. Since IL has no stabilization effect on these charge-separated intermediates, rapid [2 + 3]-addition might take place in the IL solvent.

Ionic liquids have been considered inappropriate for strong base-mediated reactions. However, recently, several examples have been reported showing the possibility of using ILs as reaction media for strong base-mediated reactions such as the Grignard reaction. Clyburne and coworkers made the first breakthrough, demonstrating that Grignard reactions took place in a mixture of phosphonium ionic liquid ($[P_{666(14)}]$ $[NTf_2]$) and tetrahydrofuran (THF) [28]. They showed that a number of reactions were possible in the phosphonium ionic liquids, including addition of carbonyl, benzyne reaction, halogenation, and coupling reactions with phenyl Grignard reagent. However, it was found that $[P_{666(14)}][NTf_2]$ had a serious drawback: isolation of the product was very difficult from this IL due to its high solubility in both ether and hexane. On the other hand, Itoh and coworkers developed $[P_{444ME}][NTf_2]$, which could stabilize the Grignard reagent by complexation with ether oxygen moiety [29].

Nagano and Hayashi reported oxidative homocoupling reaction of aromatic Grignard reagents using iron(III) chloride ($FeCl_3$) as catalyst and 1,2-dichloroethane as an oxidant [30]. Itoh and coworkers reported remarkable acceleration of iron-catalyzed homocoupling reaction of aromatic Grignard reagent using $[P_{444ME}][NTf_2]$ as a solvent (Figure 2.17) [31]. They demonstrated the homocoupling reaction of phenyl magnesium bromide **47a** (R = H, PhMgBr) in the presence of 1 mol% of $FeCl_3$ using 1,2-dichloroethane as an oxidant; the desired coupling was completed in less than 5 min at 0°C and biaryl **48a** (R = H) was obtained in excellent yield in pure $[P_{444ME}][NTf_2]$ solvent, while it took 1 h to complete the same reaction when the reaction was carried out in ether as solvent under reflux conditions. Steric bulkiness of the substituent affected the reaction greatly. It was found that 25°C was required to complete the reaction when *o*-tolyl MgBr (**47b**) was used as substrate, and the desired biaryl compound **48b** was obtained in 86% yield (Figure 2.17), though the acceleration obtained was less than that in ether solvent because 12 h under reflux conditions was required to complete the same reaction when the reaction was carried out with ether as solvent. Hayashi proposed that the key step might be the reduction step (step A) of iron(III) cation by aryl Grignard reagent or a transmetallation step (step B) on the iron as illustrated in Figure 2.17 [30]. It is expected that both reduction and transmetallation take place more easily using electron-rich Grignard reagent; the coupling reaction of more electron-rich **47a** did indeed proceed more rapidly than that of **47b** [31]. Since the reaction mixture turned black after addition of the Grignard reagent, iron nanoparticles may be produced under the reaction conditions [32], and this may contribute to completing the coupling reaction.

2.5 Prospect

Ionic liquids are now widely recognized as suitable for use in organic reactions and offer possibilities for improvement in the control of product distribution, enhanced reactivity, ease of product recovery, catalyst immobilization, and recycling. From the standpoint of green sustainable chemistry, we have shown in this chapter selected examples of synthetic reactions that realize recyclable use of catalyst in IL solvent systems and have discussed the scope of application of ILs as reaction media. One of the benefits of using ionic liquid as solvent for organic reactions is that

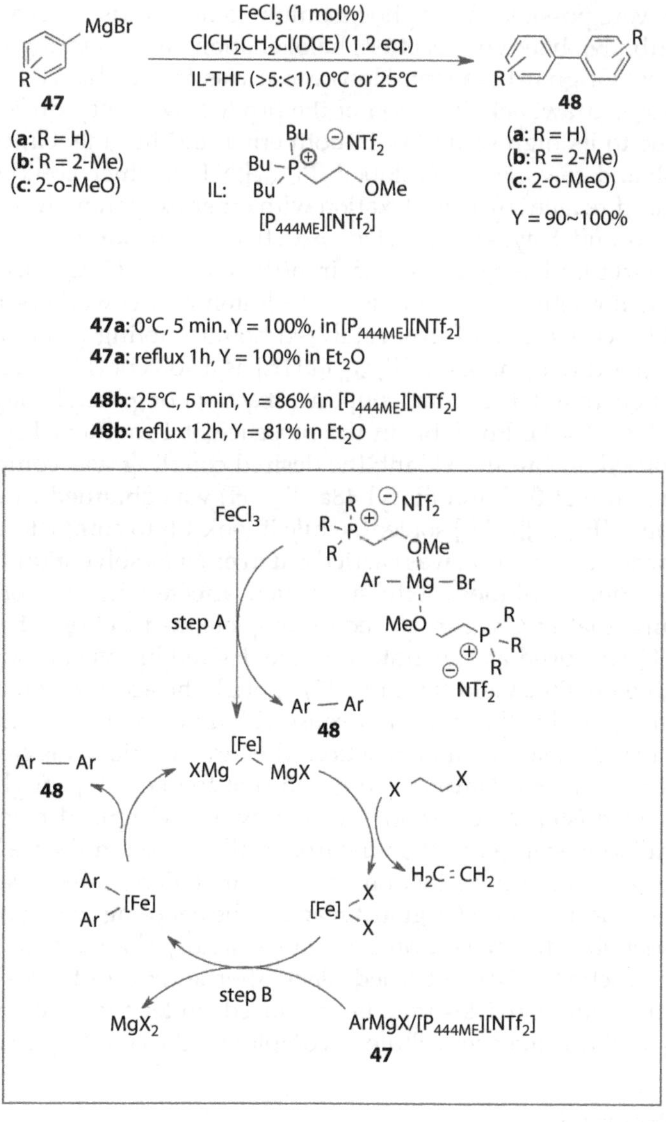

Figure 2.17 Iron(III) trichloride-catalyzed rapid homocoupling of aromatic Grignard reagent in an IL solvent system.

purification of ILs is easily accomplished by a simple method, and it can be used repeatedly. Recently, development of quick reactions has become a very important task in many stages of organic synthesis, as mentioned in the introduction. It should be emphasized that the IL solvent system has rich possibility to realize rapid chemical reactions even though the ILs are generally very viscous solvents.

References

1. Carmichael, A. J., Earle, M. J., Holbrey, J. D., McCormac, P. B., and Seddon, K. R. 1999. The Heck reaction in ionic liquids: a multiphasic catalyst system. *Org. Lett.* 1: 997–1000.
2. Welton, T. 1999. Room-temperature ionic liquids. Solvents for synthesis and catalysis. *Chem. Rev.* 99: 2071–2083.
3. Hamashima, Y., Takano, H., Hotta, D., and Sodeoka, M. 2003. Immobilization and reuse of Pd complexes in ionic liquid: efficient catalytic asymmetric fluorination and Michael reactions with β-ketoesters. *Org. Lett.* 5: 3225–3228.
4. Fujiwara, M., Kawatsura, M., Hayase, S., Nanjo, M., and Itoh, T. 2009. Iron(III) salt–catalyzed Nazarov cyclization/Michael addition of pyrrole derivatives. *Adv. Synth. Catal.* 351: 123–128.
5. Ngo, H., Hu, L. A., and Lin, W. 2005. Catalytic asymmetric hydrogenation of aromatic ketones in room temperature ionic liquid. *Tetrahedron Lett.* 46: 595–597.
6. Öchsner, E., Schneiders, K., Junge, K., Beller, M., and Wasserscheid, P. 2009. Highly enantioselective Ru-catalyzed asymmetric hydrogenation of (β)-keto ester in ionic liquid/methanol mixtures. *Appl. Catal. A Gen.* 364: 8–14.
7. Khan, H. N-ul., Prasetyanto, E. A., Kim, Y.-K., Ansari, M. B., and Park, S.-E. 2010. Chiral Cu(II) complexes as recyclable catalysts for asymmetric nitroaldol (Henry) reaction in ionic liquids as greener reaction media. *Catal. Lett.* 140:189–196.
8. Chowdari, N. S., Ramachary, D. B., and Barbas III, C. F. 2003. Organocatalysis in ionic liquids: highly efficient L-proline-catalyzed direct asymmetric Mannich reactions involving keton and aldehyde nucleophiles. *SynLett* 1906–1909.
9. Gruttadauria, M., Riela, S., Meo, P. L., D'Anna, F., and Noto, R. 2004. Supported ionic liquid asymmetric catalysis. A new method for chiral catalysts recycling. The case of proline-catalyzed aldol reaction. *Tetrahedron Lett.* 45: 6113–6116.
10. Shimakoshi, H., Baba, H., Iseki, T., Aritome, Y., Adachi, I., and Hisaeda, C. Y. 2008. Photophysical and photosensitizing properties of brominated porphycenes. *Chem. Commun.* 2882–2884.
11. Koguchi, S., and Kitazume, T. 2006. Synthetic utilities of ionic liquid-supported NHPI complex. *Tetrahedron Lett.* 47: 2797–2801.
12. Li, C., Zhao, J., Tan, R., Peng, Z., Luo, R., Peng, M., and Yin, D. 2011. Recyclable ionic liquid-bridged chiral dimeric salen Mn(III) complexes for oxidative kinetic resolution of racemic secondary alcohols. *Catal. Commun.* 15: 27–31.
13. Faber, K. 2011. *Biotransformations in Organic Chemistry: A Textbook*, 356–390. 6th ed. Springer, Heidelberg.
14. Itoh, T. 2007. Biotransformation in ionic liquid. In *Future Directions in Biocatalysis*, ed. T. Matsuda, 3–20. Elsevier Bioscience, Amsterdam.
15. van Rantwijk, F., and Sheldon, R. A. 2007. Biocatalysis in ionic liquids. *Chem. Rev.* 107: 2757–2785.
16. Itoh, T., Akasaki, E., Kudo, K., and Shirakami, S. 2001. Lipase-catalyzed enantioselective acylation in the ionic liquid solvent system: reaction of enzyme anchored to the solvent. *Chem. Lett.* 262–263.
17. (a) Amyes, T. L., Diver, S. T., Richard, J. P., Rivas, F. M., and Toth, K. 2004. Formation and stability of N-heterocyclic carbenes in water: the carbon acid pKa of imidazolium cations in aqueous solution. *J. Am. Chem. Soc.* 126,

4366–4374, 2004. (b) Magill, A. M., Cavell, K. J., and Yates, B. F. Basicity of nucleophilic carbenes in aqueous and nonaqueous solvents theoretical predictions. *J. Am. Chem. Soc.* 126, 8717–8724, 2004. (c) Tsuzuki, S., Tokuda, H., Hayamizu, K., and Watanabe, M. 2005. Magnitude and directionality of interaction in ion pairs of ionic liquids: relationship with ionic conductivity. *J. Phys. Chem. B* 109: 16474–16481.

18. Itoh, T., Nishimura, Y., Ouchi, N., and Hayase, S. 2003. 1-Butyl-2,3-dimethylimidazolium tetrafluoroborate; the most desirable ionic liquid solvent for recycling use of enzyme in lipase-catalyzed transesterification using vinyl acetate as acyl donor. *J. Mol. Catal. B Enzym.* 26: 41–45.
19. Abe, Y., Yagi, Y., Hayase, S., Kawatsura, M., and Itoh, T. 2012. Ionic liquid engineering for lipase-mediated optical resolution of secondary alcohols: design of ionic liquids applicable to ionic liquid coated-lipase catalyzed reaction. *I&EC Res.* 51: 9952–9958.
20. Cole, A. C., Jensen, J. L., Ntai, I., Tran, K. L. T., Weaver, K. J., Forbes, D. C., and Davis Jr., J. H. 2002. Novel Brønsted acidic ionic liquids and their use as dual solvent-catalysts. *J. Am. Chem. Soc.* 124: 5962–5963.
21. Earle, M. J., Seddon, K. T., Adams, C. J., and Roberts, G. 1998. Friedel–Crafts reactions in room temperature ionic liquids. *Chem. Commun.* 2097–2098.
22. Kim, H. S., Kim, Y. J., Lee, H., and Chin, C. S. 2002. Ionic liquids containing anionic selenium species: applications for the oxidative carbonylation of aniline. *Angew. Chem. Int. Ed.* 41: 4300–4303.
23. Luo, S. Z., Mi, X. L., Xu, H., Zhang, L., Liu, S., and Cheng, J.-P. 2006. Functionalized chiral ionic liquids as highly efficient asymmetric organocatalysts for Michael addition to nitroolefins. *Angew. Chem. Int. Ed.* 45: 3093–3097.
24. Kim, D. W., Song, C. E., and Chi, D. Y. 2002. New method of fluorination using potassium fluoride in ionic liquid: significantly enhanced reactivity of fluoride and improved selectivity. *J. Am. Chem. Soc.* 124: 10278–10279.
25. Kim, D. W., Ahn, D.-S., Oh, Y.-H., Lee, S., Kil, H. S., Oh, S. J., Lee, S. J., Kim, S. J., Ryu, J. S., Moon, D. H., and Chi, D. Y. 2006. A new class of SN2 reactions catalyzed by protic solvents: facile fluorination for isotopic labeling of diagnostic molecules. *J. Am. Chem. Soc.* 128: 16393–16397.
26. Kim, D. W., and Chi, D. Y. 2004. Polymer-supported ionic liquids: imidazolium salts as catalysts for nucleophilic substitution reactions including fluorinations. *Angew. Chem. Int. Ed.* 43: 483–485.
27. Ohara, H., Kiyokane, H., and Itoh, T. 2002. Cycloaddition of styrene derivatives with quinone catalyzed by ferric ion; remarkable acceleration in an ionic liquid solvent system. *Tetrahedron Lett.* 43: 3041–3044.
28. Ramnial, T., Ino, D. D., and Clyburne, J. A. C. 2005. Phosphonium ionic liquids as reaction media for strong bases. *Chem. Commun.* 325–326.
29. Itoh, T., Kude, K., Hayase, S., and Kawatsura, M. 2007. Design of ionic liquids as reaction media for the Grignard reaction. *Tetrahedron Lett.* 48: 7774–7777.
30. Nagano, T., and Hayashi, T. 2005. Iron-catalyzed oxidative homo-coupling of aryl Grignard reagents. *Org. Lett.* 7: 491–493.
31. Kude, K., Hayase, S., Kawatsura, M., and Itoh, T. 2011. Iron-catalyzed quick homocoupling reaction of aryl or alkynyl Grignard reagents using a phosphonium ionic liquid solvent system. *Heteroatom Chem.* 22: 397–404.
32. Migowski, P., and Dupont, J. 2007. Catalytic applications of metal nanoparticles in imidazolium ionic liquids. *Chem Eur J.* 13: 13–32.

chapter three

Biocatalysis in ionic liquids

Pedro Lozano, Juana M. Bernal, Eduardo García-Verdugo, Michel Vaultier, and Santiago V. Luis

Contents

3.1 Introduction: Toward a green chemical industry

The search for new environmentally benign nonaqueous solvents, which can easily be recovered/recycled, and the use of robust, selective, and efficient catalysts are two of the main goals for the development of green/sustainable chemical processes [1]. Solvents are usually used as auxiliary materials in chemical synthesis, where they act as media for mass transport, reaction, and product separation. The vast majority of solvents used in academic and industrial laboratories are molecular liquids, belonging to the group of volatile organic compounds (VOCs). They are responsible for a large part of the environmental problems of processes in the chemical industry, and have a great impact on cost, safety, and health (e.g., toxic, flammable, or corrosive compounds). Furthermore, their recovery and reuse are often associated with energy-intensive distillation and sometimes cross-contamination. The substitution or elimination of VOCs is not an easy task, because they are key elements for chemical processes (e.g., dilution of the reactants or the catalysts; to provide an environment where reactants meet, assistance for the homogeneous distribution of the energy needed for the reaction activation, or dissipation of the energy generated by an exothermal reaction; effects on performance of the catalysts; to facilitate product separation by changes in product solubility; etc.) [2].

Nowadays, ionic liquids (ILs) are the nonaqueous green solvents, also named neoteric solvents, that receive most attention worldwide. They are a new class of liquid solvents that have led to a new green chemical revolution, because of their unique array of physical-chemical properties headed by their negligible vapor pressure, which makes them suitable for numerous industrial applications [3]. However, the goal of green chemistry is much more than simply replacing hazardous solvents with environmentally benign ones. The selectivity of catalyzed processes is just as important, because of the importance to avoid undesired reactions and by-products, and facilitate products recovery. The formation of waste is also linked to the traditional use of stoichiometric amounts of reagents [4]. Switching from stoichiometric methodologies to catalytic processes is perceived as one major way to improve the efficiency of the synthetic toolbox. Thus, the greenness of chemical transformations is closely related to the use of both the catalytic [5] and the engineering [6] approach. Catalysis can improve the efficiency of a reaction by lowering the energy input required, by avoiding the use of a stoichiometric amount of reagents, and by greater product selectivity. In this regard, nature has always been a source of inspiration for chemists. To transfer the exquisite efficiency shown by enzymes in nature to chemical processes may constitute the most powerful toolbox for developing a clean and sustainable chemical industry. Enzymes, which are the catalysts of living systems with active sites designed to fit specific substrates, may well become the most suitable catalysts in chemistry since they are able to accelerate stereo-, chemo-, and regioselectively different chemical transformations, under very mild experimental conditions (low pressure and temperature, aqueous medium, etc.) [7]. They offer significant advantages with respect to catalyzed reactions by other means. The large surface contact established between a typical enzyme and its substrate, together with the well-defined active center, offers high potential for regiocontrol and stereocontrol: only one area (regiocontrol) and face (stereocontrol) of the substrate are well positioned with respect to the catalytic center, while the remainder of the molecule is excluded. As a result, a substrate molecule can often include multiple functions of similar reactivity, but only one of which will be the favored target for enzyme-catalyzed transformation (chemoselectivity) [7c].

A great variety (more than 13,000) enzyme-catalyzed reactions have been successfully demonstrated at laboratory scale, offering clear advantages for the synthesis of enantiopure fine chemicals against any other kind of catalysts [8]. Even the chemical industry is exploring the great potential of biocatalysis to manufacture both bulk and fine chemicals [9]. Furthermore, by using enzymes in nonaqueous environments, rather than in their natural aqueous reaction media, their technological applications can be greatly enhanced because of the catalytic promiscuity that results in the expansion of the repertoire of biotransformations [10]. Among the most used enzymes, lipases have gained a clear predominance, exhibiting a wide specificity to recognize very different substrates, and catalyzing different reactions used in pharmaceutical and drug production [11], biodiesel [12], or foods [13].

Over the past two decades, strategies based on engineering of reaction medium, substrate, and biocatalysts have been developed to demonstrate the suitability of enzymes as catalysts for chemical processes in nonaqueous environments [10, 14], being recently applied to both conventional organic solvents and new green nonaqueous solvents [15]. In this chapter, an overview of progress recently realized in the field of medium engineering in ionic liquids to meet the increasing demand for greener biocatalytic processes is given. Figure 3.1 summarizes the different key elements that should be considered for developing biocatalytic processes

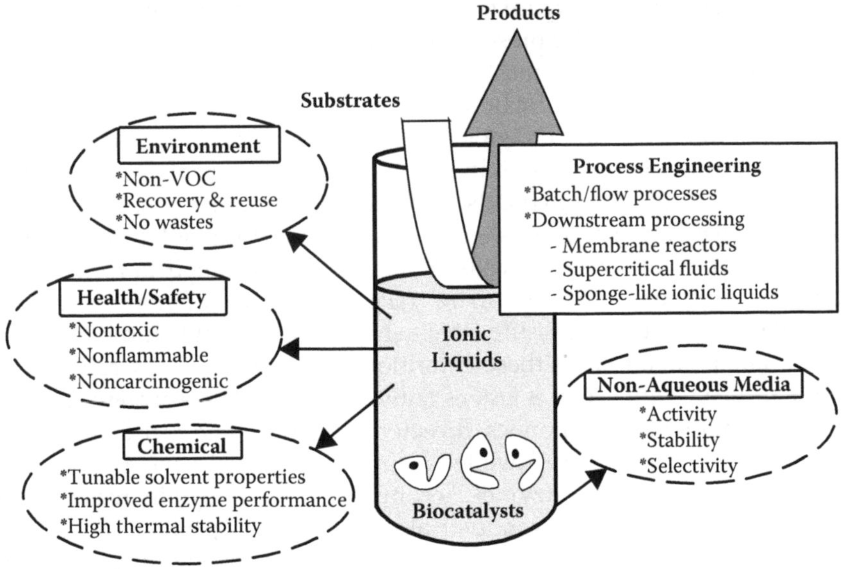

Figure 3.1 Green criteria for developing biocatalytic processes using ionic liquids.

using ionic liquids, including downstream steps or continuous operation, to push toward fully green approaches for chemical processes.

The knowledge of IL characteristics with regard to solvent properties for substrates/products, as well as the compatibility with enzymes for the maintenance or improvement of their catalytic functions, should be considered the first key task to be studied. In the same way, the suitability of reaction systems and downstream processing for developing full green chemical processes is the second key task to be addressed in this chapter. Reaction systems able to directly provide pure products, as well as the full recovery and reuse of the ILs, will be reported (e.g., membrane reactors, continuous ILs/supercritical fluid biphasic reactors, processed based on sponge-like ILs, etc.). The selected examples are mainly dealing with the use of ionic liquids for biphasic or monophasic systems, but brief insights of other systems, such as supercritical, or their combination with each other or with ILs, are also mentioned.

3.2 *Biotransformations in nonaqueous environments*

Greenness in chemical processes begins with catalysts. Enzymes, as catalysts of living systems, clearly constitute an arsenal of green catalysts for chemical processes. Enzymes are proteins, polymeric macromolecules based on amino acid units with unique sequences. They show a high level of architectonic organization, also named native conformation. Some of them require the presence of additional nonprotein components (i.e., cofactors and coenzymes) before they can carry out any catalytic function. The structure-function relationships in enzymes are key for catalytic activity, because enzyme functions strongly depend on its native conformation, which is maintained by a high number of weak internal interactions (e.g., hydrogen bonds, van der Waals interactions, etc.), as well as supramolecular interactions with other surrounding molecules, mainly water as a natural solvent of cells. Enzymes are designed to function in aqueous solutions within a narrow range of environmental conditions (e.g., pH, temperature, pressure, etc.), which in fact establish the limits of life on our planet. Out of these conditions, enzymes usually are deactivated as a consequence of a loss of native conformation through unfolding. Developments in genomics, directed evolution, and our exploitation of the natural biodiversity have led to improvements in the activity, stability, and specificity of enzymes, accompanied by a huge increase in the number and variety of their industrial applications [10, 16].

From the functional point of view, the potential of enzymes as practical catalysts cannot be doubted, since their activity and selectivity (stereo-, chemo-, and regioselectivity) for catalyzed reactions are far ranging. In this context, the market for enantiopure fine chemicals is continuously growing, making enzymes the most suitable catalysts for green synthetic processes [17]. However, the use of biocatalysts in aqueous media is limited because most chemicals of interest are insoluble in water. Furthermore, water is not chemically inert and usually gives rise to undesired side reactions. There are numerous potential advantages at employing enzymes in nonaqueous environments or reaction media with low water content (i.e., in the presence of organic solvents or additives) [10]. These advantages include the dramatically higher solubility that can be reached in the case of hydrophobic substrates, the insolubility of enzymes, which makes them easy to reuse, and the elimination of microbial contamination in reactors. Probably the most interesting advantage of using nonaqueous environments for enzyme catalysis arises when hydrolytic enzymes (e.g., lipases, esterases, proteases, glycosidases, etc.) are applied, because of the ability of these enzymes to catalyze synthetic reactions. For example, lipases have been used in synthetic organic chemistry to catalyze various kinds of reactions, such as hydrolysis, esterification, transesterification by acidolysis, transesterification by alcoholysis, interesterification, and aminolysis (see Figure 3.2a). Because of the reversibility of equilibrium, these hydrolases accelerate both the forward and the reverse reaction to the same degree, changing the rate but not the equilibrium. The position of the equilibrium alone determines whether a particular bond (e.g., ester, amide, etc.) is preferentially generated or destroyed in a given hydrolase-catalyzed reaction [18]. The synthetic mode of action needs the chemical equilibrium to be shifted toward the condensation product as a result of a drastic decrease in the water content of the reaction medium. The success of the approach is directly correlated with the reaction medium engineering used to eliminate the synthetic product or water molecules synthesized as by-products (e.g., by adding water adsorbents).

Additionally, many hydrolases (e.g., trypsin, papain, etc.) show a catalytic mechanism through a covalent acyl-enzyme intermediate as a function of the principal active group in the active site: serine (i.e., lipases, α-chymotrypsin, subtilisin, trypsin, etc.), thiol (i.e., papain), etc. In this context, synthetic reactions can be carried out easily by a kinetically controlled approach using activated substrates (i.e., vinyl esters) in nonaqueous environments, because the alcohol released, i.e., an enol, tautomerizes to a carbonyl compound (e.g., acetaldehyde), which cannot act as a substrate for the enzyme (see Figure 3.2b). The presence of an alternative nucleophile (e.g., alcohol, amine, etc.) might lead to the formation of esters or amides, respectively, rather than to the hydrolytic reaction products that occur in the presence of water.

The catalytic promiscuity of enzymes has been described in nonaqueous environments, related with the ability of a single active site to catalyze

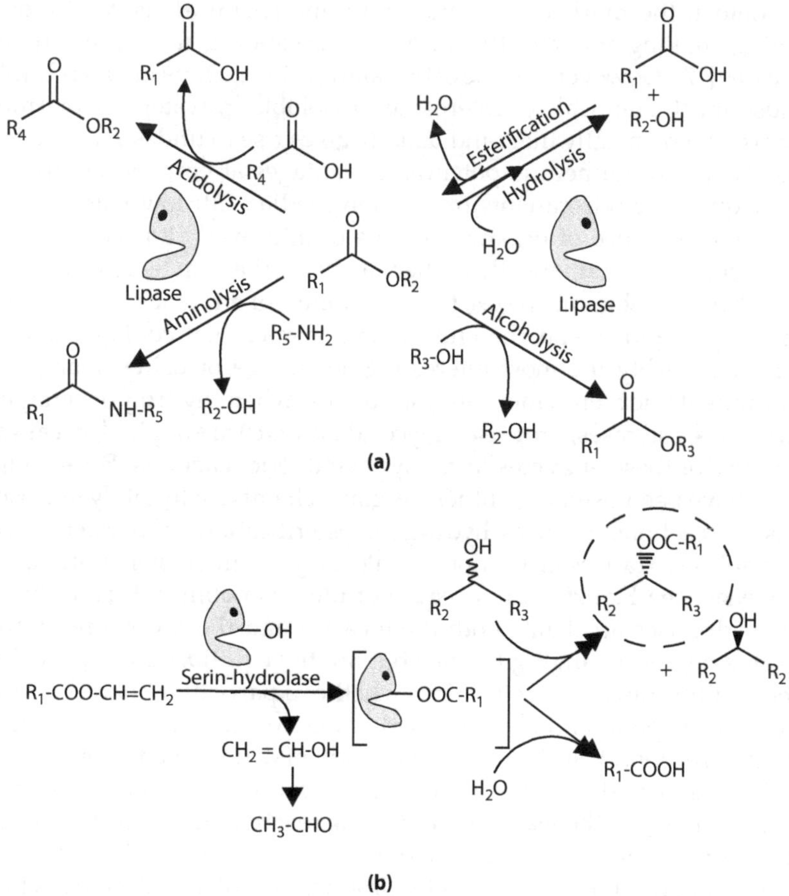

Figure 3.2 (a) Usual chemical reactions catalyzed by lipases. (b) Serin-hydrolase-catalyzed kinetic resolution of a *sec*-alcohol by using a vinyl ester as an acyl donor.

more than one chemical transformation (e.g., lipase B from *Candida antarctica* (CALB) is able to catalyze aldol additions, Michael-type additions, the formation of Si-O-Si bonds, etc.). Other enzymes (i.e., peroxidases, laccases, monooxigenases, dehydrogenases, etc.) have also been reported as being excellent tools for organic synthesis in nonaqueous environments [7, 10, 14].

However, switching from water to nonaqueous solvents, as reaction medium for enzyme-catalyzed reactions, is not always a simple answer, because the native structure of the enzyme can easily be destroyed, resulting in deactivation. Water is the key component of all nonconventional media, because of the importance that enzyme-water interactions have in maintaining the active conformation of the enzyme. Dry proteins are completely inactive, and a critical amount of water is required for them to function. It has been reported that proteins only achieve full biological activity when the surrounding water has approximately the same mass as the protein in question [19]. In this context, hydrophobic solvents typically afford higher enzymatic activity than hydrophilic ones, because the latter have a tendency to strip some of these essential water molecules from the enzyme microenvironment. Water activity (a_w) has been proposed as a key parameter for determining the correct degree of hydration of enzymes in hydrophobic organic solvents [18b], but it fails to predict the critical hydration level for enzyme activity in polar organic solvents [18c]. Enzymes typically fold in such a way that nonpolar residues are buried in a hydrophobic core, while polar residues tend to move to the surface, where they are hydrated. The common hypothesis is that when an enzyme is placed in a dry hydrophobic system, it is trapped in the native state, partly due to the low dielectric constant that greatly intensifies electrostatic forces, enabling it to maintain its catalytic activity. This is one of the most intriguing facets of nonaqueous biocatalysis, the phenomenon of the "memory" in anhydrous media [10, 18].

The compatibility of the solvent with reagents (e.g., solubility of the substrates and products, inert character, etc.) and its influence on the enzyme function (activity and stability) are important factors to be taken into account when choosing a nonaqueous medium for a given biocatalytic reaction. It is nowadays well known that enzymes can function in the absence of bulk water, that their activities can be modulated, their selectivities tailored, and their stabilities altered. However, the rates of enzyme-catalyzed reactions in nonaqueous environments are low compared to those with aqueous media, a diminution that has been attributed to several factors, such as the decrease in the polarity of the microenvironment of the enzyme, or the loss in activity during the preparation of enzymes for use in nonaqueous media (e.g., lyophilization) [20].

Enzyme immobilization is the most popular strategy that has been used to improve catalytic performance of enzymes in nonaqueous media. Supporting materials not only serve as a scaffold but also provide a unique microenvironment, which strongly affects the performance of the loaded enzyme. In this context, particular properties of supporting materials, including surface hydrophobicity/hydrophilicity, surface curvature, spatial restriction and confined microcapsule, etc., are very important criteria to prepare active and stable immobilized enzyme derivatives. Biocatalyst immobilization onto nanotubes [21] or magnetic particles [22] or hydrogels [23]), or ceramic membranes coated with hydrophilic polymers [24], etc., are recent approaches that result in the enhancement of the catalytic performance of enzymes. The use of molecular hydrogels based on a nanostructured enzyme support matrix has been reported as being an efficient system for biocatalysis in nonaqueous systems, allowing substrates and products to diffuse through a hydrophobic domain to a hydrophilic phase in which the enzyme is entrapped [23]. The amphiphilic nature of the matrix causes the hydrophilic phase to swell upon enzyme loading and the hydrophobic phase to swell in nonpolar solvents, resulting in a great increase in the turnover number of the enzyme.

Nowadays, ILs are the nonaqueous green solvents that receive the most attention worldwide for use as reaction media for enzyme-catalyzed chemical transformations, because of the improvements obtained in enzyme catalytic performances and product recovery approaches. The use of these environmentally benign nonaqueous solvents combined with efficient biocatalysts for chemical processes is strongly encouraged for developing green chemistry at an industrial scale.

3.3 *Ionic liquids and enzymes*

3.3.1 *Essential solvent properties of ionic liquids for biocatalysis*

Ionic liquids, also named room temperature ionic liquids (RTILs), molten salts, liquid organic salts, or fused salts, are a class of substances that are entirely composed of ions and which are liquid at temperatures lower than 100°C. Those liquid salts with low melting points are most desirable as solvents for reactions and processes. In contrast to conventional organic solvents, ILs have many favorable properties, including extremely low vapor pressure, a wide liquid range, low flammability, high ionic conductivity, high thermal conductivity, good dissolution power toward many substrates, high thermal and chemical stability, and wide electrochemical potential windows [4, 25]. Typical IL cations applied in biocatalysis are (1) N,N′-dialkylimidazolium, (2) tetraalkylammonium, (3) alkylpyrrolidinium, (4) N-alkylpyridinium, or (5) alkylphosphonium (see Figure 3.3), paired with a variety of anions that have a strongly delocalized negative charge, e.g., BF_4^-, PF_6^-, Tf_2N^-, etc., resulting in colorless, low viscosity, and easily handled materials [26].

Another interesting feature of ILs is the phase behavior with molecular solvents, such as organic solvents and water, because ILs can be fully miscible, partially miscible, or nonmiscible with them as a function of ions involved. Since ILs are usually composed of poorly coordinating ions, which makes them highly polar but noncoordinating solvents, they are nonmiscible with most hydrophobic organic solvents (e.g., hexane, etc.), thus providing a nonaqueous and polar media to develop two-phase systems. Also, they may be fully or partially miscible with polar organic solvents, such as ethanol, acetonitrile, etc., depending on the alkyl chain length of the cation. In the same way, most ILs are nonmiscible with water, and therefore can be used to develop biphasic systems with polar characteristics [25b]. This feature has usually been applied for developing liquid-

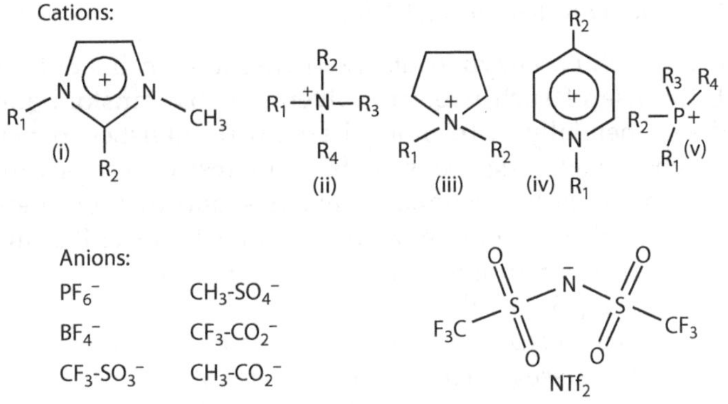

Figure 3.3 Typical structures of cations and anions involved in ILs that have been applied for enzyme catalysis. (From van Rantwijk, F., and Sheldon, R. A., *Chem. Rev.*, 107, 2757–2785, 2007.)

liquid biphasic systems (e.g., organic solvent/IL or water/IL) to recover products after a catalytic transformation in ILs [27].

The phase behavior of water/IL systems merits a special mention. By spectroscopic studies of ILs using several solvatochromic and fluorescent probes (e.g., Nile Red, Reichardt's betaine dye), it was reported how most usual ILs (e.g., [Bmim][NTf_2], [Bmim][PF_6], [Bmim][BF_4], etc.) are in the same polarity region as polar water-miscible solvents like methanol or acetonitrile [25]. Although ILs could be considered polar solvents, they cannot be considered to be similar to those polar molecular solvents with regard to their miscibility with water (e.g., ILs based on [NTf_2] or [PF_6] anions are nonmiscible with water). The water-immiscible ILs show a high hygroscopic character (e.g., [Bmim][NTf_2] is able to absorb up to 1.4% w/w water content) [28], and this feature could be regarded as an additional advantage for biocatalysis [15]. Water is the key component of biocatalytic reaction media, because of the importance of enzyme-water interactions for the maintenance of the active conformation of the enzyme. On the other hand, it should be taken into account how the presence of water in ILs based on the hexafluorophosphate anion results in the release of the undesirable hydrogen fluoride at high temperatures (>100°C) [3]. The utility of water/IL biphasic systems to recover products is limited by the IL solubility into the aqueous phase. With respect to organic solvents, developing biphasic systems with ILs for product recovery is the most used approach [26, 27], despite that it is a clear breakdown point with regard to the overall greenness of the process [15]. To overcome this problem, alternative strategies for a clean product recovery, based on the use of supercritical fluids [29], pervaporation membrane technology [30], and more recently, sponge-like ILs [31], have successfully been applied, as described at the end of this chapter.

3.3.2 Enzyme catalysis in ionic liquids

Although research on enzyme-catalyzed reactions in ILs only began in 2000 [32], the use of such neoteric solvents in biotransformations was increased exponentially, with more than 2,000 published papers today [33]. Researchers have first focused their interest on the advantages of ILs as reaction media for enzymatic catalysis, and then on understanding the exceptional behavior of enzymes in some kinds of ILs, and finally on the development of an integrated process for biotransformation and product separation [15, 26, 34].

A large number of enzymes (e.g., lipases, proteases, peroxidases, dehydrogenases, glycosidases, etc.) and reactions (e.g., esterification, kinetic resolution, reductions, oxidations, hydrolysis, etc.) have successfully been tested in ILs, because of their ability to dissolve both polar and nonpolar compounds. It is well documented from a large number of papers that an enzyme-catalyzed reaction in ILs provides better results than those obtained in conventional organic solvents: enhanced reaction rates and conversions, improved enantioselectivity and regioselectivity, etc. [24]. However, the relationships between the chemical structures of ILs and the activity, enantioselectivity, or stability displayed by enzymes are still not clearly understood.

As the biocatalyst behavior in nonaqueous media is strictly related to the degree of hydration of the protein, because of the key role of essential water molecules around the enzyme, two different reaction systems may be considered a function of the miscibility of ILs with water: aqueous solutions of ILs, and ILs in nearly anhydrous conditions.

For the case of ILs miscible with water, the activity and stability of enzymes in aqueous ILs are often discussed in terms of Hofmeister effects, being arranged according to the position of ions in these series from stabilizing to destabilizing (kosmotropic to chaotropic) [35]. However, the kosmotropic or chaotropic character of ions related does not seem to be a solid basis for predicting the compatibility of enzymes and water-miscible ILs [26a]. Aqueous solutions of ILs have been used to improve the solubility of polar substrates or products with hydrophobic moieties (e.g., amino acid derivatives), but the nature and concentration of the IL are key criteria because of the high ability of water-miscible ILs to deactivate enzymes. As a representative example, the endoprotease subtilisin is often used in the enantioselective hydrolysis of N-acylamino acid esters to S-amino acids. Organic solvents, such as acetonitrile, are added to the enzymatic system in order to increase the solubility of amino acid derivatives. It was demonstrated that the replacement of an organic solvent by 10% v/v of N-ethylpyridinium trifluoroacetate ([EPy][TFA]) IL considerably increased the activity and enantioselectivity of the reaction (see Figure 3.4), both of which fall drastically at higher concentration [36]. A

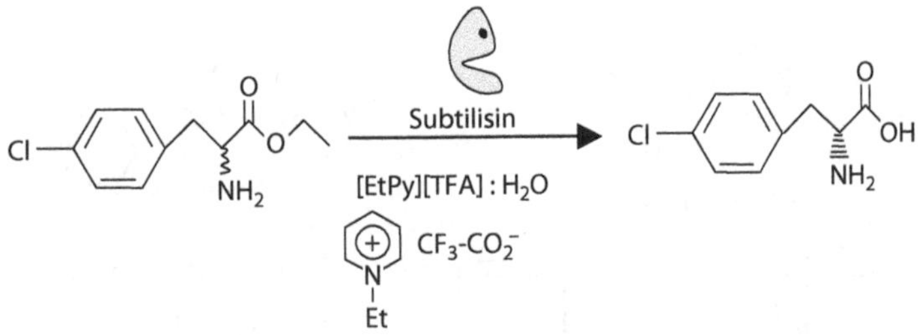

Figure 3.4 Subtilisin-catalyzed resolution of 4-chlorophenylalanine ethyl ester in [EtPy][TFA]/water as reaction medium. (From Zhao, H., and Malhotra, S. V., *Biotechnol. Lett.*, 24, 1257–1260, 2002.)

similar behavior was observed for other enzymes such as hydroxynitrile lyase, chloroperoxidase, formate dehydrogenase, β-galactosidase, etc., and attributed to the ability of ions from water-miscible ILs to deactivate enzyme by water stripping. However, some water-miscible ILs (e.g., those based on BF_4 anions or those with long alkyl chains in cations, such as cocosalkyl pentaethoxy methylammonium methosulfate ([CPMA][MS])) have been shown to be excellent reaction media for lipase-catalyzed transesterifications and glycerolysis at high IL concentration (>95% v/v) [37].

Enzymes in water-immiscible ILs have produced remarkable results at low water contents; all the assayed water-immiscible ILs (e.g., [Bmim][NTf_2], [Bmim][PF_6], [Btma][NTf_2], etc.) have been shown to act as suitable reaction media for biotransformations, appearing as viable alternatives to molecular organic solvents for organic synthesis. Lipases are by far the most used biocatalysts in water-immiscible ILs. They are used for the synthesis of aliphatic and aromatic esters, chiral esters by (dynamic) kinetic resolution of *sec*-alcohols, carbohydrate esters, polymers, etc. [26].

The usual approach applied for biotransformations consists in the direct addition of biocatalysts in the IL medium containing substrates, as illustrated in Figure 3.5 by the pioneering example of Itoh et al. [38]. As can be seen, the immobilized CALB-catalyzed kinetic resolution of 5-phenyl-1-penten-3-ol with vinyl acetate in [Bmim][PF_6] as reaction medium led to an ester yield of >99% and E-val of >500. Reaction products were extracted by liquid-liquid extraction with diethyl ether in a biphasic system, and then the excess of ether in the IL phase was eliminated under vacuum. As the enzyme remained anchored to the IL phase, this approach permits the reuse of the enzyme/IL systems by addition of fresh substrates, and without any loss in activity. In spite of liquid-liquid extraction with organic solvents being the most commonly used approach for the recovery of products after biotransformations performed in IL systems, this approach must be considered a clear breakdown with regard to the greenness of

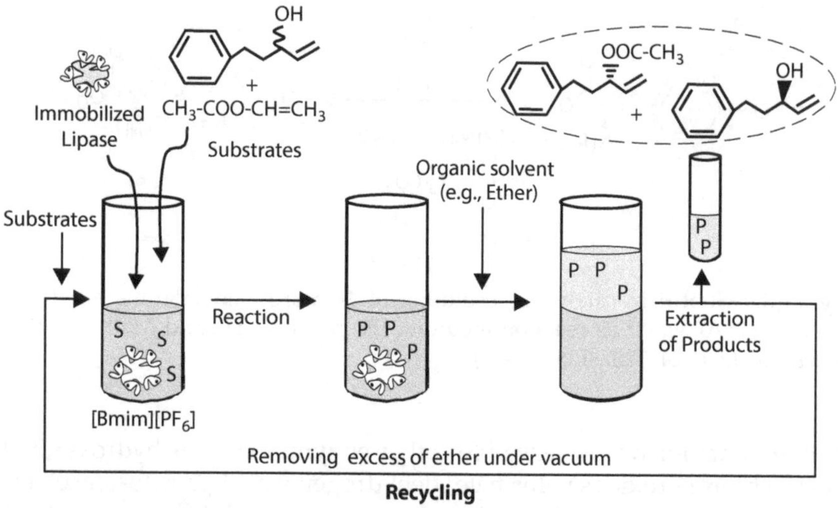

Figure 3.5 Experimental approach developed by Itoh et al. [38] for lipase-catalyzed kinetic resolution of 5-phenyl-1-penten-3-ol in a 1-butyl-3-methylimidazolium hexafluorophosphate ([Bmim][PF_6]) IL system, including recycling and product extraction steps.

the process, making it necessary to find alternative strategies for product recovery based on sustainable approaches to overcome this limitation.

The unique IL properties have also been applied to improve the performance of immobilized enzymes under nonaqueous environments by medium engineering. As an example, an efficient biocatalyst derivative based on IL-coated enzymes (ILCEs) can be prepared using a solid IL at room temperature, which melts in the range of 50–100°C. In the approach developed by Kim et al. [39], the free *Pseudomonas cepacia* lipase is stabilized by mixing the melted [1-(3′-phenylpropyl)-3-methylimidazolium)] [PF_6] IL at 53°C and then cooling the mixture and finally cutting the resulting solid ILCEs into small pieces, resulting in a useful immobilized enzyme derivative with markedly enhanced enantioselectivity and without losing any significant activity toward reuse. In the same context, the protective effect of ILs during immobilization of *Candida rugosa* lipase by the sol-gel process has been demonstrated by using hydrophobic IL [C_{16}mim][Tf_2N] as an additive [40].

Regarding the efficiency of enzymatic catalysis, it has been reported how the use of ILs improves the enzyme activity by medium engineering (e.g., synthesis of aliphatic esters [41a], synthesis of acyl L-carnitine [41b]) or the selectivity (e.g., kinetic resolution (KR) of *rac*-menthol [41c], KR of *rac*-3-phenyllactic acid [41d]) with respect to that observed in organic solvents. Classical parameters in medium engineering for biotransformations in organic solvents (e.g., aw control, pH memory, etc.) have been taken into account to improve results. In this way, salt hydrate pairs were used to control aw for CALB-catalyzed biotransformations, e.g., synthesis of 2-ethylhexyl methacrylate, synthesis of ascorbyl oleate in [Bmim][PF_6] and [Bmim][BF_4] [42], resulting in clear improvements in synthetic yields. In another example, a pH memory phenomenon at the lipase immobilization step has been described to explain the enhanced yield and enantioselectivity in IL reaction media. Magnetic microspheres containing immobilized *C. rugosa* lipase prepared via suspension polymerization were used to catalyze the resolution of *rac*-menthol in [Bmim][PF_6]. The best enantioselectivity (ee > 90%) was obtained at pH 5.0, and then decreasing gradually with increasing pH [43]. In another example, the use of 2-methoxyethyl(tri-n-butyl)phosphonium bistriflimide IL as reaction medium also resulted in an enhancement of both the lipase activity and enantioselectivity for the kinetic resolution of 4-phenylbut-3-en-2-ol, with respect to diisopropyl ether, affording the first example of a reaction rate superior to that in conventional organic solvents [44].

The most interesting biotransformations in ILs were observed at low water content or with nearly anhydrous media, but the rule for enzyme activity in ILs seems to be "there is no rule," since performance in a particular IL appears to vary significantly from enzyme to enzyme. As an example, a comparative study was carried out using free and immobilized lipases from several sources (i.e., *C. antarctica, Thermomyces lanuginosus,* or *Rhizomucor miehei*) for synthesizing butyl propionate by transesterification in anhydrous reaction media based in nine different ILs [37b]. It was observed how enzyme activities were clearly dependent on the nature of the ions, the results improving as the alkyl chain length of the imidazolium cation increased, and as a function of the type of anion ([PF_6], [BF_4], or [ethylsulfate]): the best synthetic activity was obtained when free CALB was assayed in the water-miscible IL cocosalkyl pentaethoxy methylammonium methosulfate ([CPMA][MS]). This IL was also found to be a suitable reaction medium for Novozym 435-catalyzed glycerolysis of commercial oils and fats to produce monoglycerides (90% yield), a value markedly higher than in normal solvents [45a]. The amphiphilic structure of [CPMA][MS] was suggested to be capable of creating a compatible system for glycerol, oils, and fats, as well as inducing the shift of reaction equilibrium to the formation of monoglycerides. This IL provides an excellent operational stability for the lipase, and the reusability of IL was also observed in consecutive batchwise reactions. These results were corroborated for the enzymatic production of diglycerides by the same approach using an engineered binary IL system (i.e., trioctylmethylammonium bistriflimide/[CPMA][MS]) as reaction medium [45b].

3.3.3 *Stability and stabilization of enzymes in ionic liquids*

Since the pioneering works concerning the suitability of IL reaction media for enzyme catalysis [46], the ability of these neoteric solvents to provide an environment for enhanced enzyme stability, as well as improved recyclability of biocatalysts, has widely been described [47].

Several factors of ILs, such as polarity, anion nucleophilicity, hydrogen bonding ability, excipients, impurities, pH, and the overall enzyme-substrate-medium relationship, strongly influence the stability and activity of enzymes. Enzyme stability in aqueous solutions of hydrophilic ILs has been correlated with the Hofmeister series, like inorganic salts [35], although anhydrous hydrophilic ILs provoked fast enzyme deactivation by protein unfolding. As an example, it was reported how both cellulase from *Trichoderma reesei* and serum albumins were denatured in the presence of the hydrophilic [Bmim][Cl] IL, caused by either ionic strength, water stripping, or specific binding to the protein surface, resulting in a loss in secondary structure as determined by circular dichroism (CD) [48].

However, in agreement with the suitability of water-immiscible ILs as reaction media for biotransformation processes, all proteins displayed excellent stabilities in these ILs under anhydrous conditions. In the case of free proteins (i.e., monellin, α-chymotrypsin, and CALB), the ability of water-immiscible ILs (e.g., [Bmim][NTf_2], [Btma][NTf_2], [Bmpy][NTf_2], etc.) to maintain the secondary structure and the native conformation of these proteins contrary to the usual unfolding that occurs in nonaqueous environments has been demonstrated by spectroscopic techniques (e.g., fluorescence, circular dichroism, Fourier transform infrared (FT-IR), etc.) [49]. This phenomenon has been attributed to the unique molecular characteristic of these neoteric solvents. Thus, Dupont [50] reported the structural organization of imidazolium ILs in solid and liquid phases, where these fused salts form an extended network of cations and anions connected together by hydrogen bonds (see Figure 3.6a). The monomeric unit is always constituted by one imidazolium cation surrounded by at least three anions, and in turn, each anion is surrounded by at least three imidazolium cations, where the strongest hydrogen bond always involves the most acidic H_2 of the imidazolium ring. The three-dimensional structure of these ILs is formed of chains of the imidazolium cations. In some cases, typical heterocyclic ring π/π stacking interactions also contribute to the stability of the supramolecular structure in the liquid phase. Because of this structural model, the incorporation of other molecules and macromolecules into the IL network induces changes of the three-dimensional organization of these materials and can cause, in some cases (e.g., water), the formation of polar and nonpolar regions. Wet ILs are nanostructured materials, which allow neutral molecules to reside in less polar regions and ionic or polar species to undergo faster diffusion in the more polar

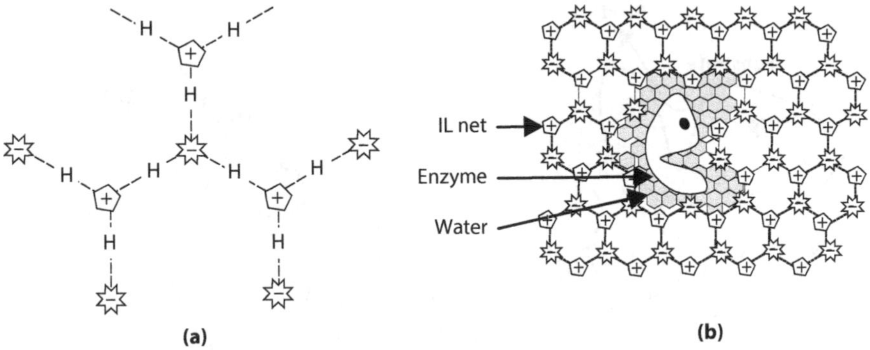

Figure 3.6 (a) Two-dimensional simplified model of the supramolecular structure of imidazolium ILs based on hydrogen bonding interactions. (b) Schematic description of the inclusion of enzymes in wet regions into the IL network.

or water-rich regions. Accordingly, enzymes in water-immiscible ILs should also be considered as included into hydrophilic gaps of the network, where the observed stabilization of enzymes may be attributed to the maintenance of this strong net around the protein (see Figure 3.6b). Furthermore, considering the protein unfolding process that occurs in water by increasing temperature, it could also be regarded as a loss of its three-dimensional structure produced by the disruption of the structure of the medium, as a consequence of the increase in the kinetic energy of water molecules when heated. The extremely ordered supramolecular structure of ILs in the liquid phase might be able to act as a mold, maintaining an active three-dimensional structure of the enzyme in aqueous nano-environments, and avoiding the classical thermal unfolding. These facts imply that free enzyme suspended in IL systems can be considered carrier-free-immobilized enzyme derivatives [15, 51].

In this context, ILs were also used as immobilizing/coating agents for enzyme stabilization in nonaqueous environments resulting in the formation of biocatalyst derivatives exhibiting better catalytic activities, stabilities, and enantioselectivities under harsh reaction conditions. For example, the coating of Novozym particles with [Bmim][NTf_2] resulted in a clear enhancement in the catalytic activity for synthesizing citronellyl alkyl esters because of the favored mass transfer phenomena around the enzyme microenvironment [52]. The development of covalently supported ionic liquid-like phases (SILLPs) either by functionalization of the styrene-divinylbenzene (PS-DVB) surfaces with IL-like (imidazolium) moieties or by polymerization of the corresponding functional monomers has open a new way to greatly reduce the amount of ILs used and to facilitate their full reuse/recovery in continuous green chemical processes. In this approach, IL properties are transferred onto the solid phase, leading to

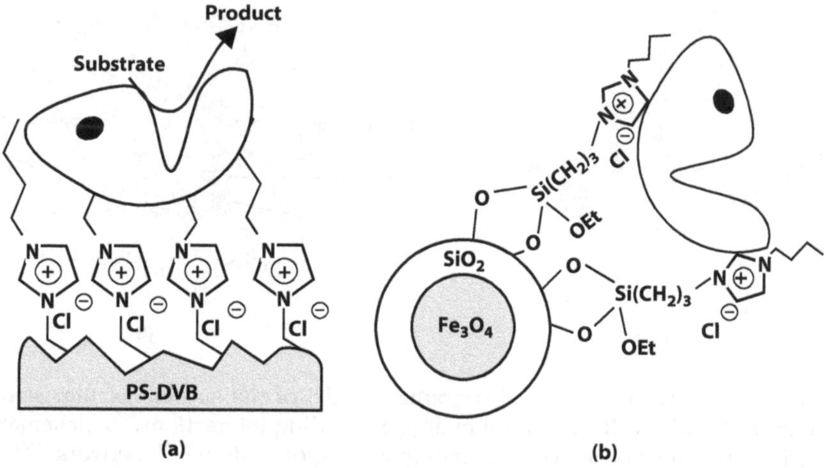

Figure 3.7 (a) Immobilized enzymes on supported ionic liquid-like phases (SILLPs) covalently attached to a PS-DVB polymeric support. (From Lozano, P. et al., *Adv. Synth. Catal.*, 349, 1077–1089, 2007.) (b) Magnetic nanoparticles. (From Jiang, Y. et al., *J. Mol. Catal. B Enzym.*, 58, 103–109, 2009.)

supported ionic liquid-like phase, either in particles or monoliths. A large diversity of SILLPs, varying cations, anions, as well as support nature and loading, have been characterized, including their thermal stability and polarity [53]. The immobilization of lipases onto SILLPs, containing imidazolium units with loadings ranging from ca. 55 to 40% wt IL per gram of polymer, resulted in highly efficient and robust heterogeneous biocatalysts for both the citronellyl propionate synthesis and the kinetic resolution of 1-phenylethanol in continuous flow [54].

In the same approach, lipase from *C. rugose* was immobilized onto magnetic nanoparticles coated by a functionalized silica shell with covalently attached IL moieties (see Figure 3.7b) [55]. The immobilized lipase derivative catalyzed the esterification of oleic acid with butanol in a solvent-free system with better activity, thermostability, and reusability than those observed for the free enzyme: immobilized lipase retained 60% of its initial activity after eight repeated batch reactions, while no activity was detected after six cycles for the free enzyme.

3.4 *Green biocatalytic processes with downstream*

In agreement with the twelve principles of green chemistry, the development of sustainable industrial processes combining the excellence of biocatalytic transformations with the unique properties of ILs necessitates appropriate reactor designs, where the full recovery and reuse of the enzyme/IL system should be demonstrated for long operation times.

Three different technological approaches have been developed for carrying out product recovery by green/sustainable techniques and avoiding the use of volatile organic solvents, as follows: membrane reactors, IL/$scCO_2$ reactors, and more recently, sponge-like IL systems.

3.4.1 *Enzymatic membrane reactor based on ILs*

Solute extraction and recovery using supported liquid membranes is recognized as one of the most promising membrane-based processes. In a supported liquid membrane system a defined solvent or solvent/carrier solution is immobilized inside the porous structure of a polymeric or ceramic membrane, in such a way that the feed phase, in which the solutes of interest are solubilized, is separated from the receiving phase, where these solutes are transferred to and, eventually, concentrated. This configuration has attracted a great deal of interest, but its industrial application has been hindered by the difficulty in designing supported liquid membranes that exhibit high stability during operation [56].

The use of ILs as an immobilized phase in a supporting membrane is particularly interesting due to the nonvolatile character of these neoteric solvents and their solubility properties in the surrounding phases, which makes it possible to obtain very stable supported liquid membranes without any observable loss of the IL to the atmosphere or the contacting phases. In this way, the feasibility of using ILs as a new kind of solvent in supported liquid membranes for selective transport of organic molecules was reported, being studied in different mixtures of compounds with representative organic functional groups (e.g., hexylamine, methylmorpholine, etc.). The appropriate combination of selected ILs and supporting membranes was crucial for achieving good selectivity in a given separation problem [57].

This approach is of high interest for the continuous highly selective separation of mixtures of isomeric organic compounds that are structurally similar and have close boiling points, by using the ILs (e.g., [Bmim][PF_6]) in a supported liquid membrane (SLM). The SLM permits the selective separation of a target molecule by exploiting the solubility differences between solutes in the liquid membrane phase, as demonstrated for the kinetic resolution of the *rac*-ibuprofen case (see Figure 3.8a) [58]. The membrane reactor system was operated by coupling two lipase reactions (esterification and hydrolysis, respectively) with a membrane containing a supported IL. In the system, each lipase was placed at each side of the membrane and the IL-based supported liquid membranes permit the selective transport of organic molecules; the system provides for easy and selective permeation of the synthesized S-ibuprofen ester through the membrane. In the receiving aqueous phase, the ester is then hydrolyzed by another lipase that provides a successful resolution of the racemic mixture.

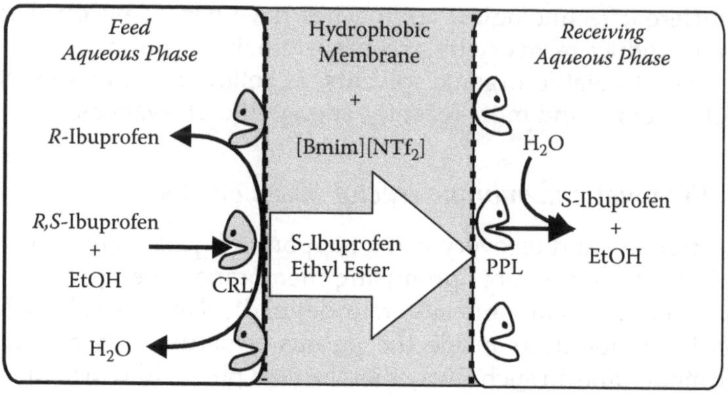

Figure 3.8 Enantioselective transport of S-ibuprofen through a supported IL membrane with two immobilized lipase: *C. rugosa* lipase (CRL) and porcine pancreas lipase (PPL). (From Belafi-Bako, K. et al., *Desalination*, 149, 2002, 267–268.)

Membrane reactors based on pervaporation systems have also been developed to facilitate separation of volatile products, or to shift equilibrium reactions in which water is formed as one of the products, affecting negatively both substrate conversion and enzyme activity for the catalyzed esterification reaction. The pervaporation technique is defined as a selective transport of liquid through a homogeneous nonporous membrane with simultaneous evaporation of permeates. It can be coupled with a reactor unit and in such form enable selective removal of volatile compounds. During pervaporation, the feed mixture is contacted with the active side of the nonporous membrane, and the components passed through the membrane are recovered as vapor and the secondary side is usually under high vacuum [59]. This approach has been used for the enzymatic synthesis of flavor esters (e.g., ethyl acetate, isoamyl acetate) by esterification in ionic liquid reaction media [60]. As can be seen in Figure 3.9, the immobilized CALB-catalyzed esterification of acetic acid with isoamyl alcohol occurred into a stirred-tank reactor containing $[Bmim][PF_6]$ as reaction medium. The coupling of two membrane units on the reactor permitted removing of both the isoamyl acetate and water products by a double pervaporation system using hydrophobic and hydrophilic membranes, respectively, in continuous operation for 72 h without any loss in the enzyme activity [60a].

3.4.2 *Bioprocesses in IL/supercritical carbon dioxide biphasic systems*

A supercritical fluid (SCF) is defined as a state of matter at a pressure and temperature higher than its critical point, but below the pressure required to condense it into a solid. SCFs are characterized by gas-like viscosities

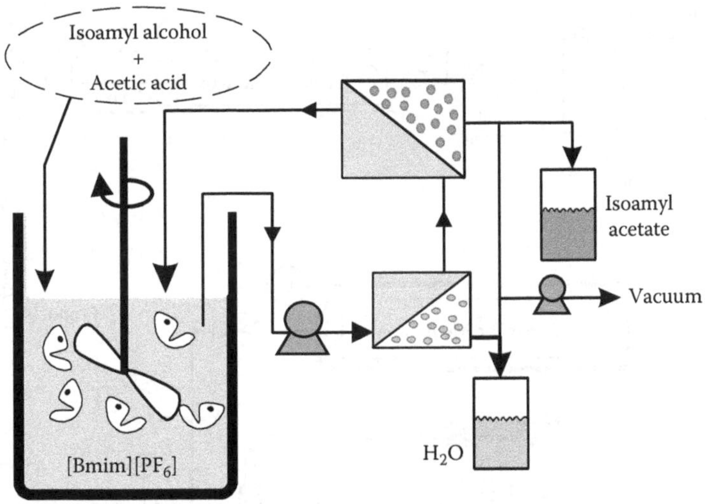

Figure 3.9 Novozym 435-catalyzed isoamyl acetate synthesis in [Bmim][PF_6] using a stirred tank reactor coupled to pervaporation units for product separation. (From Feher, E. et al., *Desalination*, 241, 8–13, 2009.)

and solvating properties of a wide range of various organic solvents and derivatives. Physical properties of these solvents are unique because they may be tuned simply by adjusting the pressure and temperature, showing exceptional abilities for extraction, reaction, fractionation, and analysis processes. The key feature of SCFs is the sensitivity of the fluid density to both pressure and temperature, especially in the critical point vicinity [61]. Between all SCFs, the environmentally benign supercritical carbon dioxide ($scCO_2$) is the most popular. Its great potential to develop cleaner alternative processes has been demonstrated. Products can be easily freed from solvent traces, which is highly attractive as an alternative reaction medium for food and pharmaceutical products. The $scCO_2$ is chemically inert, nontoxic, nonflammable, cheap, readily available, and shows relatively low critical parameters (e.g., Pc = 73.8 bar, Tc = 31.0°C), remaining by far the most popular supercritical solvent for (bio)catalysis, as described in other chapters.

Biocatalytic processes in $scCO_2$ are usually performed in high-pressure vessels by both discontinuous and continuous flow operations [62]. Immobilized enzymes onto solid supports can be used in a packed bed reactor, leading to a simple continuous enzyme reuse analogous to consecutive cycles without depressurization requirements. Reaction parameters, such as temperature, pressure, reaction dilution, flow rates, etc., are experimental variables used to optimize a synthetic transformation in a continuous biocatalytic process in $scCO_2$. As an example, the production of long-chain fatty acid esters (e.g., alkyl oleate) is of great interest

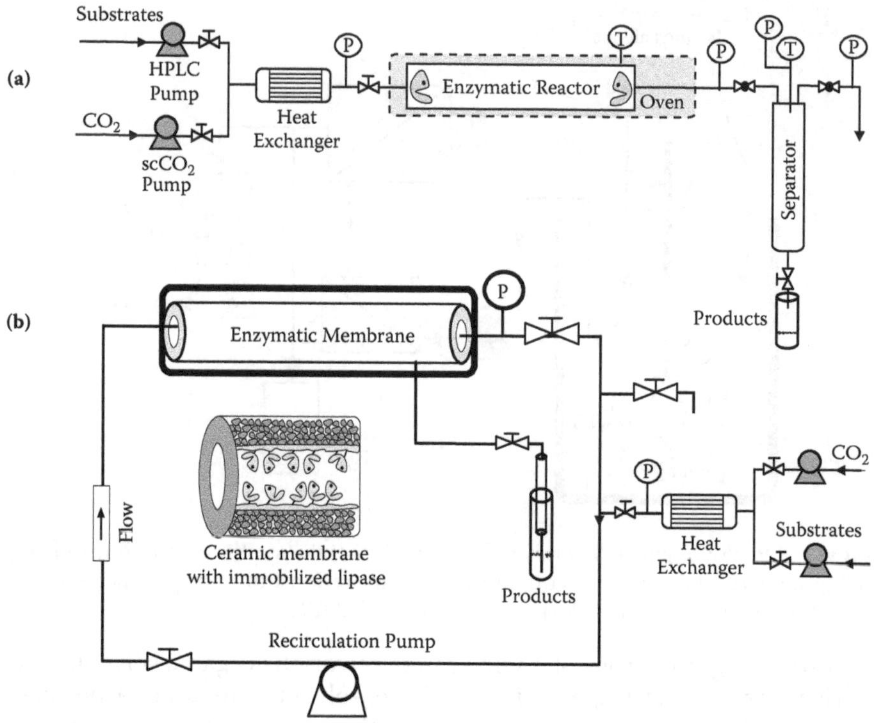

Figure 3.10 (a) Setup of continuous flow biocatalytic reactor for synthesis of long-chain fatty acid esters in $scCO_2$. (From Laudani, C. G., *J. Supercrit. Fluids,* 41, 74–81, 2007.) (b) High-pressure membrane reactor with recirculation for enzyme-catalyzed butyl butyrate synthesis in $scCO_2$. (From Lozano, P. et al., *J. Supercrit. Fluids,* 29, 2004, 121–128.)

for the cosmetic, pharmaceutical, and lubricant industries. Thus, the esterification of oleic acid with 1-octanol (Figure 3.3a) catalyzed by *R. miehei* lipase immobilized on a macroporous anion exchange resin (Lipozyme RM) has been studied. In a first example, this reaction was performed in batch and continuous packed bed reactors using dense CO_2 as solvents (Figure 3.10a), allowing a reach up to 93% conversion under the optimized conditions, with a productivity maintained for a long-term period without any significant reduction [63]. Enzymatic membrane reactors have also been assayed as an alternative to a packed bed for continuous biocatalytic processes in $scCO_2$ [64]. These membrane reactors constitute an attempt to combine catalytic conversion, product separation or concentration, and catalyst recovery into a single operation. As an example, enzymatic dynamic membranes are formed by depositing water-soluble polymers (e.g., gelatine, polyethyleneimine, etc.) on a ceramic porous support. This alternative reactor is a porous membrane contactor along and through which substrates are continuously flowing. It was first applied for CALB-catalyzed butyl butyrate synthesis in a semicontinuous mode under supercritical conditions (Figure 3.10b). The reactor was operated in daily cycles (6 h of a continuous synthetic process in the selected conditions and 18 h of storage in the reactor at room temperature), showing an excellent operational behavior, without practically any loss of activity during the assayed time (half-life time higher than 360 cycles). The better enzymatic activity exhibited by the dynamic membrane in $scCO_2$ with respect to the organic solvents clearly showed the necessity of a relevant selection of an experimental conditions set and reactor design in order to avoid the possible adverse effects of CO_2 on enzyme activity [24a, 65]. The feasibility of this new process was demonstrated on a cross-flow filtration unit operating under $scCO_2$ conditions, showing high activity and stability of immobilized lipases.

In this context, it is necessary to highlight how the classical advantages of $scCO_2$ to extract, dissolve, and transport chemicals are tarnished in enzymatic processes because of its denaturative effect on enzymes. The chemically inert character of CO_2 could be in doubt regarding its interaction with proteins. The CO_2 forms carbamates with ε-amino groups of lysine residues placed on the enzyme surface, and decreases the pH of the aqueous layer around the enzyme. Both phenomena have been directly related with the usual enzyme deactivation observed in $scCO_2$ [66]. In addition to these specific effects of CO_2 on proteins, the high pressure may also have a negative impact on enzyme conformation. The rapid release of CO_2 dissolved in the bound water of the enzyme during depressurization has been claimed to produce structural changes in the enzyme and to cause its inactivation [67]. Thus, the review published by Beckman and Russel in 1999 [62a] states in the final sentence: "the advantages of replacing conventional organic solvents with supercritical fluids have not been fully demonstrated yet."

In this context, the use of IL/$scCO_2$ biphasic systems as reaction media for enzyme catalysis has opened up new opportunities for integral green process development in nonaqueous environments, because it permits us to combine the enzyme-protective effect of ILs and the classical advantages of $scCO_2$ to extract, dissolve, and transport chemicals. The pioneering work of Brennecke's group in 1999 showed that ILs (e.g., [Bmim][PF_6]) and $scCO_2$ form biphasic systems [29a]. Additionally, although $scCO_2$ is highly soluble in the IL phase and is able to extract previously dissolved hydrophobic compounds (e.g., naphthalene), the same IL is not measurably soluble in the $scCO_2$ phase. This discovery was crucial for further developments in multiphase green (bio)catalytic processes involving both chemical transformation and extraction steps [15, 68].

Multiphase biocatalytic systems based on ILs and $scCO_2$ were originally described in 2002, and were the first operational approach for the

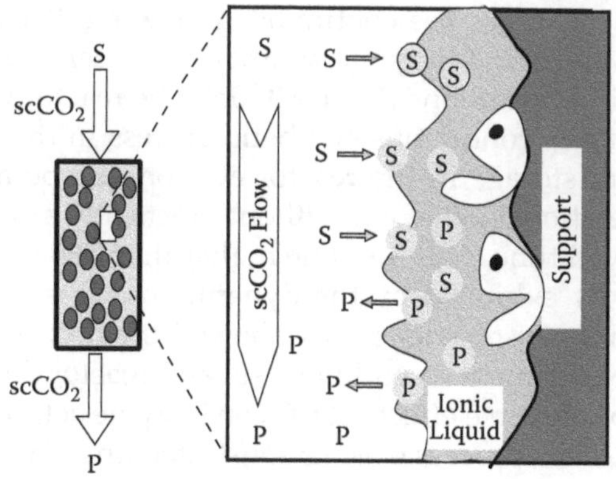

Figure 3.11 Schema of a continuous green enzymatic reactor working in ILs/ $scCO_2$ biphasic systems. (From Lozano, P. et al., *Chem. Commum.*, 2002, 692–693.)

development of fully green chemical processes in nonaqueous environments [70]. Using this approach, the $scCO_2$ flow can serve both to transport the substrate to the IL phase containing the biocatalyst and to extract the product(s) from the IL phase. Subsequently, products are obtained free from IL and from other organic solvent residues by SCF decompression, whereas CO_2 can be recycled by recompression. Additionally, if the reaction product does not require any further purification, the approach enhances the economic benefit of the process, because the system runs as a black box able to transform pure substrates into pure products without waste generation. By using this approach, continuous green biphasic biocatalytic systems in nonaqueous environments have been designed by dissolving free enzymes into the IL and then adsorbed onto a solid support, or by coating supported enzyme molecules with ILs. Biotransformations then occur in an IL phase (catalytic phase), while substrates and products remain largely in the SCF phase (extractive phase) (Figure 3.11).

By this approach, two different reactions catalyzed by CALB were tested, i.e., aliphatic ester synthesis by transesterification between 1-alkanols and vinyl esters (e.g., butyl butyrate from vinyl butyrate and 1-butanol), and the kinetic resolution of *rac*-1-phenylethanol in a wide range of conditions (100–150 bar and 40–100°C). Under these conditions, the enzyme showed an exceptional level of activity, enantioselectivity (ee > 99.9), and operational stability (e.g., the enzyme only lost 15% activity after 11 cycles of 4 h) [69a]. Thus, excellent results can also be obtained for biotransformations in $scCO_2$ using the enzyme coated with ILs even under extreme conditions, such as 100 bar and 150°C [70], as well as to understand the role of the IL phase on mass transport phenomena between $scCO_2$ and the IL phase as a function of the ion's nature [71]. Other studies have pointed to the importance of the acyl donor nature in kinetic resolution reactions to permit the selective separation of enantiomeric product (i.e., 66% yield, ee > 99.9%) by connecting reactor with cryo-traps at the appropriate pressure and temperature [72].

Integrated multicatalytic processes, whereby one initial substrate is catalytically transformed into one final product by two or more consecutive catalytic steps in the same reaction system, is of great interest for developing future chemical industry [73]. Dynamic kinetic resolution (DKR) processes can be taken as an example to illustrate a multicatalytic approach in IL/scCO_2 biphasic systems. A DKR process is based on the combination of an enzymatic kinetic resolution (KR) reaction with an in situ chemical racemization of the undesired enantiomer, which theoretically should permit reach up to 100% of one enantiomeric pure product.

The continuous DKR processes of *rac*-1-phenylethanol in IL/scCO_2 biphasic systems were first carried out combining immobilized CALB with silica modified with benzenosulfonic acid groups as catalysts in a packed bed reactor under scCO_2 at 50°C and 100 bar. Both chemical and enzymatic catalysts were previously coated with ILs (e.g., [Emim][NTf_2], [Btma][NTf_2], or [Bmim][PF_6]) at a 1:1 (w:w) ratio to prevent enzyme deactivation by scCO_2 [74]. A simple mixture of both catalysts resulted in a complete loss of activity, probably due to the acid environment around the enzyme particles. However, catalyst particles packed under three different layers (immobilized enzyme/acid catalyst/immobilized enzyme) physically separated by glass wool, led to encouraging results for the *R*-ester product (76% yield, 91–98% ee). For this reactor configuration, the *R*-ester product yield may tend to 100% only if several enzymatic and acid catalyst layers are stacked in the packed bed, according to a dichotomic progression. It is also worth noticing how the presence of the undesired *S*-ester and hydrolytic products in the scCO_2 flow was enhanced when the acid catalyst particles were assayed without IL coating.

The use of weaker solid acids, such as zeolites, as chemical catalyst clearly improved the efficiency of the continuous DKR of *rac*-1-phenylethanol in combination with immobilized CALB under scCO_2 flow (see Figure 3.12a) [75]. The best results (98% yield, 96% ee) were obtained for a heterogeneous mixture between faujasite-type zeolite (CBV400) particles coated with [Btma][NTf_2] and Novozym 435 particles coated with the same IL. Due to the low acidity of the assayed zeolites, the packaging of the heterogeneous mixture of catalyst particles coated with IL did not result in any activity loss of the immobilized CALB during 14 days of continuous operation in CO_2 under different supercritical conditions. This work clearly demonstrated the exciting potential of multicatalytic (enzymatic or chemoenzymatic) systems in ILs/scCO_2 for synthesizing optically active pharmaceutical drugs by a sustainable approach. A further

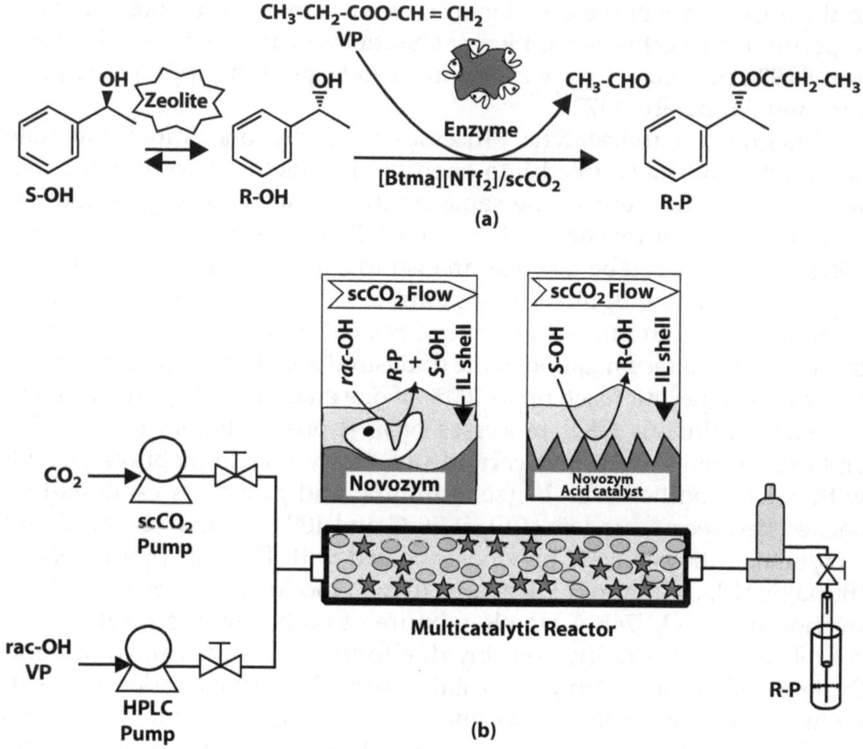

Figure 3.12 (a) DKR of *sec*-alcohols (*rac*-OH) catalyzed by the combined action immobilized enzyme (Novozym 435) and an acid zeolite chemical catalyst. (b) Setup of a continuous packed bed reactor containing both immobilized lipase and acid catalyst coated with [Btma][NTf$_2$]. (From Lozano, P. et al., *Green Chem.*, 11, 2009, 538–542.)

step toward an optimized system arose from the development and application of lipase-SILLP combinations.

In another example pushing forward the interest of IL/scCO$_2$ biphasic systems, CALB-SILLP derivatives were combined with zeolites (i.e., CP811E-150) for carrying out the DKR of the phenylethanol by using a single reactor in a continuous way. In this reactor, both the immobilized enzyme and zeolite were mixed together to perform a continuous one-pot catalytic minireactor in scCO$_2$. The selectivity of the chemical catalysts was improved by coating with a small amount of a hydrophobic IL (i.e., [BMIM][PF$_6$]). The efficient combination of CALB-SILLP and zeolites-IL resulted in increased yields of the desired *R*-product up to 92% with ee of >99.9% for the continuous DKR of *rac*-phenylethanol with vinyl propionate in scCO$_2$ [54b].

3.4.3 Bioprocesses in sponge-like ILs

Between all the arrays of unique properties of ILs as neoteric solvents, a new term has recently been coined, i.e., sponge-like ILs. This refers to hydrophobic ILs with long alkyl side chains that behave as temperature switchable ionic liquid/solid phases. These ILs have been shown as suitable reaction media for carrying out integrated clean biocatalytic approaches of synthesis and pure product separation. It was reported how ILs based on cations with long alkyl side chains, such as N-octadecyl-N′,N″,N‴-trimethylammonium bis(trifluoromethylsulfonyl)imide ([C_{18}tma][NTf_2]), 1-ocadecyl-3-methylimidazolium bis(trifluoromethylsulfonyl)imide, ([C_{18}mim][NTf_2]), etc., having melting points higher than room temperature are able to dissolve hydrophobic compounds (i.e., triolein, geraniol, citronellol, etc.) in the liquid phase, providing excellent monophasic reaction media for the lipase-catalyzed transesterification or esterification reactions with industrial interest, such as the enzymatic synthesis of flavor esters [31], and the biocatalytic synthesis of biodiesel [76].

Flavor esters of short-chain carboxylic acid (e.g., geranyl propionate, isoamyl acetate, etc.) are important fragrance compounds used in the food, cosmetic, and pharmaceutical industries. They can be labeled as "natural" when prepared by physical processes (e.g., extraction from natural sources) or by biotransformation of precursors isolated from nature [77]. Enzyme-catalyzed direct esterification of natural substrates in solvent-free and anhydrous media could be considered the most straightforward way to produce natural flavor esters. However, moderate yields (e.g., maximum of 60% isoamyl acetate) and the deactivation of enzyme by direct contact with acids are severe drawbacks for industrial application [78]. In this context, the biocatalytic synthesis of 16 different flavor alkyl esters by direct esterification of a alkyl carboxylic acid (acetic, propionic, butyric, or valeric) with a flavor alcohol (citronellol, geraniol, nerol, or isoamyl alcohol) was reported in different ILs (e.g., ([C_{16}tma][NTf_2, etc.) as switchable ionic liquid/solid phases, reaching product yields near 100% after 4 h at 50°C. The catalytic activity of the enzyme/IL system was shown to be practically unchanged after reuse [31].

Furthermore, it was observed how these resulting fully clear monophasic reaction media, containing both the flavor product and the IL, became monophasic solid systems after cooling to room temperature. These solid phases could be separated into two phases, an upper liquid containing the flavor ester and another bottom solid containing the IL by following an iterative centrifugation protocol and lowering the temperature to 4°C. The upper phase was an IL-free flavor ester up to 0.757 g/ml concentration, as determined by ^{19}F nuclear magnetic resonance (NMR) studies (see Figure 3.13a). This unique feature was explained as a function

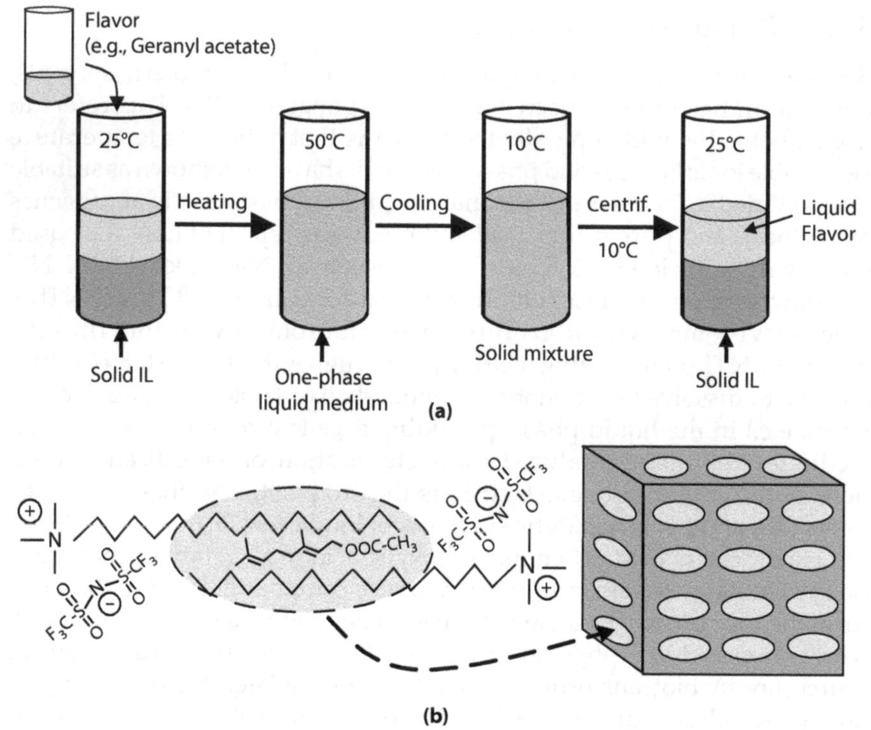

Figure 3.13 (a) Scheme of the phase behavior of flavor ester/[C_{16}tma][NTf_2] mixtures as a function of temperature. (b) Schematic hypothesis of the sponge-like organization of solid [C_{16}tma][NTf_2] net with hydrophobic holes containing liquid geranyl acetate. (From Lozano, P. et al., *Green Chem.*, 14, 2012, 3026–3033.)

of a sponge-like behavior of these temperature switchable ionic liquid/solid phases, where flavor esters were considered as being included rather than dissolved in the liquid/solid IL phases. The decrease in free volume of the ionic net produced by cooling allowed compaction of the IL solid phase by centrifugation, and the consequent release of geranyl acetate molecules outside the net. Thus, the IL net could be considered a nanosponge with holes of variable volume, which are suitable for housing or releasing hydrophobic molecules as a function of their liquid or solid phase, respectively (see Figure 3.13b).

The production of biodiesel by the biocatalytic approach is another example of a clean process in sponge-like IL. Biodiesel is usually synthesized by transesterification of triacylglycerides with methanol, yielding fatty acid methyl esters (FAMEs) and glycerol as by-products. Three consecutive transesterification reactions are involved in the full conversion of triacylglyceride molecules to biodiesel using chemical or enzymatic catalysts and an excess of alcohol to shift the equilibrium toward the product side. Although biodiesel is produced on an industrial scale using alkaline catalysts like KOH, the undesired side reactions (e.g., soap formation), glycerol recovery, and removal of inorganic salts remain important economical and environmental problems. The use of enzymes as catalysts for biodiesel synthesis is regarded as the perfect solution to these problems because of their high catalytic activity and selectivity. However, the nonmiscibility of the triglycerides and the methanol substrates is the key feature that limits the catalytic efficiency of any transesterification process to synthesize biodiesel [12]. The resulting vegetable oil/methanol biphasic systems strongly reduce reaction rates, and for the case of biocatalysis, this also causes full and fast enzyme deactivation as a result of the direct interaction between the catalytic protein and the methanol phase. In this context, it was reported how ILs based on cations with long alkyl side chains (e.g., [C_{18}mim][NTf_2], [C_{18}tma][NTf_2], etc.) were able to dissolve both triolein and methanol at any concentration, providing one-phase reaction media that showed excellent suitability for the biocatalytic synthesis of biodiesel, i.e., up to 96% yield in 6 h at 60°C. Furthermore, the excellent suitability of the enzyme/IL system for biodiesel synthesis was also demonstrated by the total preservation of the catalytic activity for subsequent reuse [79]. Furthermore, it was also observed how the reaction mixture becomes solid by cooling, allowing it to easily be fractionated by iterative centrifugations at controlled temperature into three phases: solid IL, glycerol, and pure biodiesel. The sponge-like ILs were able to "soak up" biodiesel as a liquid phase, and then could be "wrung out" by centrifugation in the solid phase [76]. These results represent a straightforward and sustainable approach for producing biodiesel, allowing the full recovery and reuse of the biocatalyst/IL system for successive cycles and suitable for scaling up, opening up a new way in green chemistry for separating products from reaction media based on ILs.

3.5 *Prospects*

The transfer of the smart catalytic efficiency displayed by enzymes in living systems to chemical processes may constitute the most powerful toolbox for developing a clean and sustainable chemical industry in the near future. The technological applications of biocatalysts are enhanced in nonaqueous environments, because of the resulting expansion of the repertoire of enzyme-catalyzed transformations, where the selective synthesis of chiral products of high commercial value is of the highest interest. Furthermore, these exquisite catalytic properties of enzymes are improved by using ILs as reaction media. Because of their exceptional ability to overstabilize enzymes even under extremely harsh conditions, ILs bring an important added value to the construction of a sustainable chemical industry. Besides, the unique properties of ILs, which can be tailored at the molecular level by an appropriate selection of the cation, the length and nature of the alkyl chain attached, or the anion, have opened a new window on processing options not available using conventional organic solvents.

The unique properties of hydrophobic ILs, based on cations with long alkyl side chains (e.g., [C_{18}tma][NTf_2]), as switchable ionic liquid/solid phases, can be extended to include a new feature: their behavior as sponge-like systems. This fact opens up a new way in green chemistry for separating products from reaction media based on soak-up and wrung-out phenomena of these sponge-like ILs. These ILs can soak up hydrophobic compounds as a liquid phase, which became a single solid phase by cooling. Then, the solid phase behaves like a sponge from which the compounds of interest are recovered as a pure liquid phase by centrifugation of the solid, like a wrung-out phenomenon. This opens up a new avenue in green chemistry for separating products from reaction media based on ILs. In the same context, $scCO_2$ seems to be the perfect companion of ILs for the development of downstream steps in green synthetic processes, which allows cleaning and recovering of ILs for reuse because of the unique phase behavior of ILs/$scCO_2$ systems. By combining enzymes with ILs/$scCO_2$ biphasic systems, the chemical industry has another clear strategy for developing integral green synthetic processes.

A key element to successfully implement all these catalytic methodologies in chemical industry is the integration of catalytic steps in multistep organic syntheses (i.e., multienzymatic or multichemoenzymatic reactor) in sponge-like ILs or IL/$scCO_2$ biphasic systems, mimicking the metabolic pathways found in nature. The novelty of the approaches described in this chapter should help open the door for developing the green chemical industry of the near future.

Acknowledgments

We thank MINECO, Spain (CTQ2011–28903-C02–02), and SENECA Foundation, Spain (08616/PI/08), for support.

References

1. (a) P.T. Anastas, M. M. Kirchhoff, T. C. Williamson. Catalysis as a foundational pillar of green chemistry. *Appl. Catal. A*. 2001, 221, 3–13; (b) C.R. Reichardt. Solvents and solvent effects: an introduction. *Org. Process Res. Develop*. 2007, 11, 105–113; (c) C.J. Li and B.M. Trost. Green chemistry for chemical synthesis. *Proc. Natl. Acad. Sci. U. S. A*., 2008, 105, 13197–13202; (d) R.A. Sheldon. E factors, green chemistry and catalysis: An odyssey. *Chem. Commun*. 2008, 3352–3365; (e) C. J. Li, P. T. Anastas. Green Chemistry: present and future. *Chem Soc Rev*. 2012, 41, 1413–1414.

2. (a) D.J.C. Constable, A.D. Curzons, V. L. Cunningham. Metrics to "green" chemistry—which are the best? *Green. Chem.* 2002, 4, 521–527; (b) E. Buncel, R. Stairs, H. Wilson. *The Role of the Solvents in Chemical Reactions*. Oxford University Press: Oxford, 2003.
3. (a) P. Wasserscheid and T. Welton (Eds.). *Ionic Liquids in Synthesis*. Wiley-VCH: Weinheim, 2008; (b) H. Zhao, S. V. Malhotra. Applications of ionic liquids in organic synthesis. *Aldrichim. Acta*. 2002, 35, 75–83; (c) R. D. Rogers, K. R. Seddon. Ionic liquids—Solvents of the future? *Science* 2003, 302, 792–793; (d) J. Dupont. From molten salts to ionic liquids: A "nano" journey. *Acc. Chem. Res.* 2011, 44, 1223–1231.
4. (a) P.T. Anastas, L.G. Heine, T.C. Williamson (Eds.). *Green Chemical Syntheses and Processes*. ACS Publications, 2001; (b) R.A. Sheldon. Fundamentals of green chemistry: efficiency in reaction design. *Chem. Soc. Rev.* 2012, 41, 1437–1451.
5. (a) R.A. Sheldon, I. Arends, U. Hanefeld (Eds.). *Green Chemistry and Catalysis*. Wiley-VCH Verlag GmbH & Co. KGaA: Weinheim, 2007; (b) G. Rothenberg (Ed.). *Catalysis, Concepts and Green Chemistry*. Wiley-VCH Verlag GmbH: Weinheim, 2008.
6. (a) J.H. Clark, D. Macquarie (Eds.). *Handbook of Green Chemistry & Technology*. Blackwell Science Ltd: Oxford, 2002; (b) R. L. Lankey, P.T. Anastas (Eds.). *Advancing Sustainability through Green Chemistry and Engineering*. ACS Publications, 2005; (c) S. Wener, M. Haumann, P. Wasserscheid. Ionic liquids in chemical engineering. *Annu. Rev. Chem. Biomol. Eng.* 2010, 1, 203–230.
7. (a) A.S. Bommarius, B.R. Riebel. (Eds.). *Biocatalysis*. Wiley-VCH Verlag GmbH & Co. KGaA: Weinheim, 2004. (b) P.T. Anastas, R. Crabtree. (Eds.). *Handbook of Green Chemistry-Green Catalysis: Biocatalysis*, vol. 3. Wiley-VCH Verlag GmbH: New York, 2009; (c) C. M. Clouthier, J. N. Pelletier. Expanding the organic toolbox: a guide to integrating biocatalysis in synthesis. *Chem. Soc. Rev.* 2012, 41, 1585–1605; (d) M.J. Liszka, M.E. Clark, E. Schneider, D.S. Clark. Nature versus nurture: Developing enzymes that function under extreme conditions. *Annu. Rev. Chem. Biomol. Eng.* 2012, 3, 77–102.
8. (a) M. Sureshkumar, C. K. Lee. Biocatalytic reactions in hydrophobic ionic liquids. *J. Mol. Catal. B: Enzym.* 2009, 60, 1–12; (b) T. Hudlicky, J.W. Reed. Applications of biotransformations and biocatalysis to complexity generation in organic synthesis. *Chem. Soc. Rev.* 2009, 38, 3117–3132; (c) E. Barbayianni, G. Kokotos. Biocatalyzed regio- and chemoselective ester cleavage: synthesis of bioactive molecules. *ChemCatChem*. 2012, 4, 592–608.
9. (a) S.M. Roberts, G. Casy, M.-B. Nielsen, S. Phythian, C. Todd, K. Wiggins. *Biocatalysts for Fine Chemicals Synthesis*. John Wiley & Sons, 1999; (b) A. Liese, K Seelbach, C. Wandrey (Eds.). *Industrial Biotransformations: A Comprehensive Handbook*. Wiley-VCH: Weinheim, 2006.
10. (a) A.M. Klibanov. Improving enzymes by using them in organic solvents. *Nature* 2001, 409, 241–246; (b) H. L. Yu, L Ou, J. H. Xu. New trends in non-aqueous biocatalysis. *Curr. Org. Synth.* 2010, 19, 1424–1432; (c) R.C. Rodrigues, R. Fernandez-Lafuente. Lipase from *Rhizomucor miehei* as an industrial biocatalyst in chemical process. *J. Mol. Catal. B Enzym.* 2010, 64, 1–22; (d) M.S. Humble, P. Berglund. Biocatalytic promiscuity. *Eur. J. Org. Chem.* 2011, 19, 3391–3401; (e) M. Kapoor, G. N. Gupta. Lipase promiscuity and its biochemical applications. *Process Biochem.* 2012, 47, 555–569.

11. (a) F. Hasan, A. Shah, A. Hameed. Industrial applications of microbial lipases. *Enzyme Microb. Technol.* 2006, 39, 235–251; (b) J. Gonzalez-Sabin, R. Moran-Ramallal, F. Rebolledo. Regioselective enzymatic acylation of complex natural products: expanding molecular diversity. *Chem. Soc. Rev.* 2011, 40, 5321–5335; (c) R. Kourist, P.D. de Maria, K. Miyamoto. Biocatalytic strategies for the asymmetric synthesis of profens—recent trends and developments. *Green Chem.* 2011, 13, 2607–2618.
12. J. M. Bernal, P. Lozano, E. Garcia-Verdugo, M.I. Burguete, G. Sanchez-Gomez, G. Lopez-Lopez, M. Pucheault, M. Vaultier, S.V. Luis. Supercritical synthesis of biodiesel. *Molecules* 2012, 17, 8696–8719; (b) A. Gog, M. Roman, M. Tosa, C. Paizs, F. Irime. Biodiesel production using enzymatic transesterification—current state and perspectives. *Renew. Energ.* 2012, 39, 10–16.
13. (a) R. C. Rodrigues, R. Fernandez-Lafuente. Lipase from *Rhizomucor miehei* as a biocatalyst in fats and oils modification. *J. Mol. Catal. B Enzym.* 2010, 66, 15–32; (b) R. Fernandez-Lafuente. Lipase from *Thermomyces lanuginosus*: uses and prospects as an industrial biocatalyst *J. Mol. Catal. B Enzym.* 2010, 62, 197–212.
14. (a) M. N. Gupta, I. Roy. Enzymes in organic media—forms, functions, and applications. *Eur. J. Biochem.* 2004, *271*, 2575–2583; (b) E. Busto, V. Gotor-Fernandez, V. Gotor. Hydrolases: catalytically promiscuous enzymes for non-conventional reactions in organic synthesis. *Chem. Soc. Rev.* 2010, 39, 4504–4523; (c) A. Rioz-Martinez, G. de Gonzalo, D. E. T. Gonzalo, M. W. Fraaije, V. Gotor. Enzymatic synthesis of novel chiral sulfoxides employing Baeyer–Villiger monooxygenases. *Eur. J. Org. Chem.* 2010, 33, 6409–6416.
15. P. Lozano. Enzymes in neoteric solvents: from one-phase to multiphase systems. *Green Chem.* 2010, 12, 555–569.
16. (a) R. N. Patel. Enzymatic synthesis of chiral intermediates for drug development. *Adv. Synth. Catal.* 2001, 343, 527–546; (b) M. T. Reetz. Directed evolution of enantioselective enzymes as catalysts for organic synthesis. *Adv. Catal.* 2006, 49, 1–69; (c) M. Wang, T. Si, H. Zhao. Biocatalyst development by directed evolution. *Bioresour. Technol.* 2012, 115, 117–125.
17. (a) A. Ghanem. Trends in lipase-catalyzed asymmetric access to enantiomerically pure/enriched compounds. *Tetrahedron* 2007, 63, 1721–1754; (b) J. H. Tao, J.H. Xu. Biocatalysis in development of green pharmaceutical processes. *Curr. Opin. Chem. Biol.* 2009, 13, 43–50.
18. (a) A. Fitzpatrick, A. M. Klibanov. How can the solvent affect enzyme enantioselectivity. *J. Am. Chem. Soc,* 1991, 113, 3166–3171; (b) P. J. Halling. Thermodynamic predictions for biocatalysis in nonconventional media—theory, test, and recommendations for experimental design and analysis. *Enzyme Microb. Technol.* 1994, 16, 178–206; (c) G. Bell, A. E. M. Janssen, P. H. Halling. Water activity fails to predict critical hydration level for enzyme activity in polar organic solvents: interconversion of water concentrations and activities. *Enzyme Microb. Technol.* 1997, 20, 471–477.
19. H. Frauenfelder, G. Chen, J. Berendzen, P.W. Fenimore, H. Jansson, B.H. McMahon, I.R. Stroe, J. Swenson, R.D. Young. A unified model of protein dynamics. *Proc. Natl. Acad. Sci. USA* 2009, 106, 5129–5134.
20. (a) P.A. Fitzpatrick, A.C.U. Steinmetz, D. Ringe, A.M. Klibanov. Enzyme crystal-structure in a neat organic solvent. *Proc. Natl. Acad. Sci. USA* 1993, 90, 8653–8657; (b) M.Y. Lee, J.S. Dordick. Enzyme activation for nonaqueous media. *Curr. Opin. Biotechnol.* 2002, 13, 376–384; (c) P. Lozano, T. De Diego, J.L. Iborra.

Immobilization of enzymes for use in supercritical fluid. In J. M. Guisan (Ed.), *Immobilization of Enzymes and Cells: Methods in Biotechnology Series*, vol. 22, pp. 269–282. Humana Press Inc.: Totowa, 2006; (d) R.A. Sheldon. Enzyme immobilization: the quest for optimum performance. *Adv. Synth. Catal.* 2007, 349, 1289–1307; (e) F. Gao, G. Ma. Effects of microenvironment on supported enzymes. *Top. Catal.* 2012, 55, 1114–1123.

21. (a) P. J. Ji, H. S. Tan, X. Xu, W. Feng. Lipase covalently attached to multiwalled carbon nanotubes as an efficient catalyst in organic solvent. *AICHE J.* 2010, 56, 3005–3011; (b) H. S. Tan, W. Feng, P.J. Ji. Lipase immobilized on magnetic multi-walled carbon nanotubes. *Bioresour. Technol.* 2012, 115, 172–176.
22. E.T. Hwang, B. Lee, M. Zhang, S. H. Jun, J. Shim, J. Lee, J. Kim, M. B. Gu. Immobilization and stabilization of subtilisin Carlsberg in magnetically-separable mesoporous silica for transesterification in an organic solvent. *Green Chem.* 2012, 14, 1884–1887; (b) M. P. Marszall, T. Siodmiak. Immobilization of *Candida rugosa* lipase onto magnetic beads for kinetic resolution of (R,S)-ibuprofen. *Catal. Commun.* 2012, 24, 80–84.
23. (a) N. Bruns, J.C. Tiller. Amphiphilic network as nanoreactor for enzymes in organic solvents. *Nano Lett.* 2005, 5, 45–48; (b) Q. Wang, Z. Yang, L. Wang, M. Ma, B. Xu. Molecular hydrogel-immobilized enzymes exhibit superactivity and high stability in organic solvents. *Chem. Commun.* 2007, 1032–1034; (c) J. Le Bideau, L. Viau, A. Vioux. Ionogels, ionic liquid based hybrid materials *Chem. Soc. Rev.* 2011, 40, 907–925; (d) K. Gawlitza, C. Z. Wu, R. Georgieva, D.Y. Wang, M.B. Ansorge-Schumacher, R. von Klitzing. Immobilization of lipase B within micron-sized poly-N-isopropylacrylamide hydrogel particles by solvent exchange. *Phys. Chem. Chem. Phys.* 2012, 14, 9594–9600;
24. (a) P. Lozano, G. Villora, D. Gómez, A.B. Gayo, J.A. Sánchez-Conesa, M. Rubio, J.L. Iborra. Membrane reactor with immobilized *Candida antarctica* lipase B for ester synthesis in supercritical carbon dioxide. *J. Supercrit. Fluids* 2004, 29, 121–128; (b) M. Mori, R. G. Garcia, M. P. Belleville, D. Paolucci-Jeanjean, J. Sanchez, P. Lozano, M. Vaultier, G.M. Rios. A new way to conduct enzymatic synthesis in an active membrane using ionic liquids as catalyst support. *Catal. Today* 2005, 104, 313–317.
25. (a) J. Dupont, R.F. de Souza, P.A.Z. Suarez. Ionic liquid (molten salt) phase organometallic catalysis. *Chem. Rev.* 2002, 102, 3667–369; (b) C.F. Poole. Chromatographic and spectroscopic methods for the determination of solvent properties of room temperature ionic liquids. *J. Chromatrogr. A* 2004, 1037, 49–82; (b) C. Reichardt. Pyridinium N-phenoxide betaine dyes and their application to the determination of solvent polarities part 29—polarity of ionic liquids determined empirically by means of solvatochromic pyridinium N-phenolate betaine dyes. *Green Chem.* 2005, 7, 339–351.
26. (a) F. van Rantwijk, R.A. Sheldon. Biocatalysis in ionic liquids *Chem. Rev.* 2007, 107, 2757–2785; (b) C. Roosen, P. Muller, L. Greiner. Ionic liquids in biotechnology: applications and perspectives for biotransformations. *Appl Microbiol. Biotechnol.* 2008, 81, 607–614; (c) T. Itoh. Recent development of enzymatic reaction systems using ionic liquids. *J. Synth. Org. Chem. Japan* 2009, 67, 143–155; (d) M. Moniruzzaman, N. Kamiya, M. Goto. Activation and stabilization of enzymes in ionic liquids. *Org. Biomolec. Res.* 2010, 8, 2887–2899; (e) M. Naushad, Z. A. Alothman, A. B. Khan, M. Ali. Effect of ionic liquid on activity, stability, and structure of enzymes: a review. *Int. J. Biol. Macromol.* 2012, 51, 555–560.

27. S. Oppermann, F. Stein, U. Kragl. Ionic liquids for two-phase systems and their application for purification, extraction, and biocatalysis *Appl. Microbiol. Biotechnol.* 2011, 89, 493–499.
28. P. Bonhote, A.P. Dias, N. Papageorgiou, K. Kalyanasundaram, M. Gra1tzel. Hydrophobic, highly conductive ambient-temperature molten salts. *Inorg. Chem.* 1996, 35, 1168–1178.
29. (a) L.A. Blanchard, D. Hancu, E.J. Beckman, J.F. Brennecke. Green processing using ionic liquids, and CO_2. *Nature* 1999, 399, 28–29; (b) S. Keskin, D. Kayrak-Talay, U. Akman, O. Hortaçsu. A review of ionic liquids towards supercritical fluid applications. *J. Supercrit. Fluids* 2007, 43, 150–180; (c) M. Roth. Partitioning behaviour of organic compounds between ionic liquids and supercritical fluids. *J. Chromatogr. A* 2009, 1216, 1861–1880.
30. E. Feher, V. Illeova, I. Kelemen-Horvath, K. Belafi-Bako, M. Polakovic, L. Gubicza. Enzymatic production of isoamyl acetate in an ionic liquid–alcohol biphasic system. *J. Mol. Catal. B: Enzym.* 2008, 50, 28–32; (b) L. Gubicza, K. Belafi-Bako, E.Feher, T. Frater. Waste-free process for continuous flow enzymatic esterification using a double pervaporation system *Green Chem.* 2008, 10, 1284–1287.
31. P. Lozano, J.M. Bernal, A. Navarro. A clean enzymatic process for producing flavour esters by direct esterification in switchable ionic liquid/solid phases. *Green Chem.* 2012, 14, 3026–3033.
32. M. Erbeldinger, A. J. Mesiano, A. J. Russell. Enzymatic catalysis of formation of Z-aspartame in ionic liquid—an alternative to enzymatic catalysis in organic solvents. *Biotechnol. Prog.* 2000, 16, 1129–1131; R. Madeira-Lau, F. van Rantwijk, K. R. Seddon, R. A. Sheldon. Lipase-catalyzed reactions in ionic liquids. *Org Lett.* 2000, 2, 4189–4191.
33. ISI Web of Knowledge, search. Topic: (ionic liquid* AND biocatal*) OR (ionic liquid* AND enzym*). Year published: 2000–2012.
34. S. Cantone, U. Hanefeld, A. Basso. Biocatalysis in non-conventional media-ionic liquids, supercritical fluids, and the gas. *Green Chem.* 2007, 9, 954–971; P. Dominguez. "Nonsolvent" applications of ionic liquids in biotransformations and organocatalysis. *Angew. Chem. Int. Ed.* 2008, 47, 6960–6968; Y. X. Fan, J. Q. Qian. Lipase catalysis in ionic liquids/supercritical carbon dioxide and its applications. J. *Mol. Catal. B Enzym.* 2010, 66, 1–7; P. Lozano and E. García-Verdugo. Non-conventional solvents in biotransformations: medium engineering and process development. In P. Dominguez de Maria (Ed.), *Ionic Liquids in Biotransformations and Organocatalysis: Solvents and Beyond*, pp 103–150. Wiley-VCH. 2012.
35. (a) H. Zhao. Are ionic liquids kosmotropic or chaotropic? An evaluation of available thermodynamic parameters for quantifying the ion kosmotropicity of ionic liquids. *J. Chem. Technol. Biotechnol.* 2006, 81, 877–891; (b) H. Zhao, O. Olubajo, Z. Y. Song, A. L. Sims, T. E. Person, R. L. Lawal, L. A. Holley. Effect of kosmotropicity of ionic liquids on the enzyme stability in aqueous solutions. *Bioorg. Chem.* 2006, 34, 15–25; (c) H. Weingartner, C. Cabrele, C. Herrmann. How ionic liquids can help to stabilize native proteins. *Phys. Chem. Chem. Phys.* 2012, 14, 415–426
36. H. Zhao, S. V. Malhotra. Enzymatic resolution of amino acid esters using ionic liquid N-ethyl pyridinium trifluoroacetate. *Biotechnol Lett.* 2002, 24, 1257–1260.

37. (a) Z. Guo, X. Xu. Lipase-catalyzed glycerolysis of fats and oils in ionic liquids: a further study on the reaction system *Green Chem.* 2006, 8, 54–62; (b) T. De Diego, P. Lozano, M.A. Abad, K. Steffensky, M. Vaultier, J.L. Iborra. On the nature of ionic liquids and their effects on lipases that catalyze ester synthesis. *J. Biotechnol.* 2009, 140, 234–241.
38. T. Itoh, E. Akasaki, Y. Nishimura. Efficient lipase-catalyzed enantioselective acylation under reduced pressure conditions in an ionic liquid solvent system *Chem. Lett.* 2002, 154–155.
39. (a) J.K. Lee, M.J. Kim. Ionic liquid-coated enzyme for biocatalysis in organic solvent *J. Org. Chem.* 2002, 67, 6845–6847; (b) M.J. Kim, J.K. Lee. Enzymes coated with ionic liquid. US Patent 7,005,282, 2006.
40. (a) S.H. Lee, T.T.N. Doan, S.H. Ha, W.J. Chang, Y.M. Koo. Influence of ionic liquids as additives on sol-gel immobilized lipase. *J. Mol. Catal. B: Enzym* 2007, 47, 129–134; (b) S.H. Lee, T.T. Doan, S. H. Ha, Y.M. Koo. Using ionic liquids to stabilize lipase within sol-gel derived silica *J. Mol. Catal. B* 2007, 45, 57–61.
41. (a) P. Lozano, T. De Diego, D. Carrie, M. Vaultier, J.L. Iborra. Enzymatic ester synthesis in ionic liquids. *J. Mol Catal. B: Enzym.* 2003, 21, 9–13; (b) J. Q. Tian, Q. Wang, Z. Y. Zhang. Lipase-catalyzed acylation of l-carnitine with conjugated linoleic acid in $[Bmim]PF_6$ ionic liquid. *Eur. Food Res. Technol.* 2009, 229, 357–363; (c) D. H. Zhang, S. Bai, M. Y. Ren, Y. Sun, *Food Chem.* 2008, 109, 72–80; (d) L. Banoth, M. Singh, A. Tekewe, U. C. Banerjee. Increased enantioselectivity of lipase in the transesterification of dl-(+/-)-3-phenyllactic acid in ionic liquids. *Biocatal. Biotransf.* 2009, 27, 263–270.
42. (a) J. A. Berberich, J. L. Kaar, A. J. Russell. Use of salt hydrate pairs to control water activity for enzyme catalysis in ionic liquids. *Biotechnol. Prog.* 2003, 19, 1029–1032; (b) M. Adamczak, U. T. Bornscheuer. Improving ascorbyl oleate synthesis catalyzed by *Candida antarctica* lipase B in ionic liquids and water activity control by salt hydrates. *Process Biochem.* 2009, 44, 257–261.
43. M. Y. Ren, S. Bai, D. H. Zhang, Y. Sun. pH memory of immobilized lipase for (+/-)-menthol resolution in ionic liquid. *J. Agric. Food Chem.* 2008, 56, 2388–2391.
44. Y. Abe, K. Kude, S. Hayase, M. Kawatsura, K. Tsunashima, T. Itoh. Design of phosphonium ionic liquids for lipase-catalyzed transesterification. *J. Mol. Catal. B: Enzym.* 2008, 51, 81–85.
45. (a) Z. Guo, X.B. Xu. New opportunity for enzymatic modification of fats and oils with industrial potentials. *Org. Biomol. Chem.* 2005, 3, 2615–2619; (b) Z. Guo, D. Kahveci, B. Ozcelik, X.B. Xu. Improving enzymatic production of diglycerides by engineering binary ionic liquid medium system. *New Biotechnol.* 2009, 26, 37–43.
46. P. Lozano, T. De Diego, D. Carrie, M. Vaultier, J.L. Iborra. Over-stabilization of *Candida antarctica* lipase B by ionic liquids in ester synthesis. *Biotechnol. Lett.* 2001, 23, 1529–1533.
47. (a) M. Persson, U. T. Bornscheuer. Increased stability of an esterase from Bacillus stearothermophilus in ionic liquids as compared to organic solventes. *J. Mol. Catal. B: Enzym.* 2003, 22, 21–27; (b) J.L. Kaar, A.M. Jesionowski, J.A. Berberich, R. Moulton, A. J. Russell. Impact of ionic liquid physical properties on lipase activity and stability. *J. Am. Chem. Soc.* 2003, 125, 4125–4131; (c) O. Ulbert, K. Belafi-Bako, K. Tonova, L. Gubicza. Thermal stability enhancement of *Candida rugosa* lipase using ionic liquids. *Biocatal. Biotrans.*

2005, 23, 177–183; (d) E. Feher, B. Major, K. Belafi-Bako, L. Gubicza. On the background of enhanced stability and reusability of enzymes in ionic liquids. *Biochem. Soc. Trans*. 2007, 35, 1624–1627.
48. M.B.Turner, S.K. Spear, J.G. Huddleston, J.D. Holbrey, R.D. Rogers. Ionic liquid salt-induced inactivation and unfolding of cellulase from *Trichoderma reesei*. *Green Chem*. 2003, 5, 443–447.
49. (a) S.N. Baker, T.M. McCleskey, S. Pandey, G.A. Baker. Fluorescence studies of protein thermostability in ionic liquids. *Chem. Commun*, 2004, 940–941; (b) T. De Diego, P. Lozano, S. Gmouh, M. Vaultier, J.L. Iborra. Fluorescence and CD spectroscopic analysis of the alpha-chymotrypsin stabilization by the ionic liquid, 1-ethyl-3-methylimidazolium bis[(trifluoromethyl)sulfonyl] amide. *Biotechnol. Bioeng*. 2004, 88, 916–924; (c) T. De Diego, P. Lozano, S. Gmouh, M. Vaultier, J.L. Iborra. Understanding structure—stability relationships of *Candida antartica* lipase B in ionic liquids. *Biomacromolecules* 2005, 6, 1457–1464; (d) F. van Rantwijk, F. Secundo, R.A. Sheldon. Structure and activity of *Candida antarctica* lipase B in ionic liquids. *Green Chem*. 2006, 8, 282–286; (e) W.Y. Lou, M.H. Zong, T.J. Smith, H. Wu, J.F. Wang. Impact of ionic liquids on papain: an investigation of structure–function relationships. *Green Chem*. 2006, 8, 509–512.
50. J. Dupont. On the solid, liquid and solution structural organization of imidazolium ionic liquids *J. Braz. Chem. Soc*. 2004, 15, 341–350.
51. P. Lozano, T. De Diego, S. Gmouh, M. Vaultier, J.L. Iborra. Dynamic structure–function relationships in enzyme stabilization by ionic liquids. *Biocatal. Biotransf*. 2005, 23, 169–176.
52. P. Lozano, R. Piamtongkam, K. Kohns, T. De Diego, M. Vaultier, J.L. Iborra. Ionic liquids improve citronellyl ester synthesis catalyzed by immobilized *Candida antarctica* lipase B in solvent-free media. *Green Chem*. 2007, 8, 780–784.
53. M.I. Burguete, F. Galindo, E. Garcia-Verdugo, N. Karbass, S.V. Luis. Polymer supported ionic liquid phases (SILPs) versus ionic liquids (ILs): how much do they look alike? *Chem. Commun*. 2007, 3086–3088.
54. (a) P. Lozano, E. Garcia-Verdugo, R. Piamtongkam, N. Karbass, T. De Diego, M.I. Burguete, S.V. Luis, J.L. Iborra. Bioreactors based on monolith-supported ionic liquid phase for enzyme catalysis in supercritical carbon dioxide. *Adv. Synth. Catal*. 2007, 349, 1077–1084; (b) P. Lozano, E. García-Verdugo, N. Karbass, K. Montague, T. De Diego, M.I. Burguete, S.V. Luis. Supported ionic liquid-like phases (SILLPs) for enzymatic processes: continuous KR and DKR in SILLP-scCO_2 systems. *Green Chem*. 2010, 12, 1803–1810; (c) P. Lozano, E. García-Verdugo, J. M. Bernal, D.F. Izquierdo, M.I. Burguete, G. Sánchez-Gómez, S.V. Luis. Immobilised lipase on structured supports containing covalently attached ionic liquids for the continuous synthesis of biodiesel in scCO_2. *ChemSusChem* 2012, 5, 790–798.
55. Y. Jiang, C. Guo, H. Xia, I. Mahmood, C. Liu and H. Liu. Magnetic nanoparticles supported ionic liquids for lipase immobilization: enzyme activity in catalyzing esterification. *J. Mol. Catal. B: Enzym*. 2009, 58, 103–109.
56. G.M. Rios, M.P. Belleville, D. Paolucci-Jeanjean. Membrane engineering in biotechnology: quo vamus? *Trends. Biotechnol*. 2007, 25, 242–246.
57. L.C. Branco, J.G. Crespo, C.A.M. Afonso. Studies on the selective transport of organic compounds by using ionic liquids as novel supported liquid membranes. *Chem. Eur. J*. 2002, 8, 3866–3871; L.C. Branco, J.G. Crespo, C.A.M.

Afonso. Highly selective transport of organic compounds by using supported liquid membranes based on ionic liquids. *Angew. Chem. Int. Ed.* 2002, 41, 2771–2773.

58. (a) E. Miyako, T. Maruyama, N. Kamiya, M. Goto. Enzyme-facilitated enantioselective transport of (S)-ibuprofen through a supported liquid membrane based on ionic liquids. *Chem. Commun.* 2003, 2926; (b) E. Miyako, T. Maruyama, N. Kamiya, M. Goto. A supported liquid membrane encapsulating a surfactant-lipase complex for the selective separation of organic acids. *Chem. Eur. J.* 2005, 11, 1163–1170.
59. Z. Findrik, G. Nemeth, D. Vasic-Racki, K. Belafi-Bako, Z. Csanadi, L. Gubicza. Pervaporation-aided enzymatic esterifications in non-conventional media. *Process Biochem.* 2012, 47, 1715–1722.
60. K. Belafi-Bako, N. Dormo, O. Ulbert, L. Gubicza. Application of pervaporation for removal of water produced during enzymatic esterification in ionic liquids. *Desalination* 2002, 149, 267–268.
61. (a) Brunner, G. (Ed.). *Supercritical Fluids as Solvents and Reaction Media.* Elsevier BV, 2004; (b) P.G. Jessop, T. Ikariya, R. Noyori. Homogeneous catalysis in supercritical fluids. *Chem. Rev.* 1999, 99, 475–493; (c) C.M. Rayner. The potential of carbon dioxide in synthetic organic chemistry. *Org. Proc. Res. Develop.* 2007, 11, 121–132.
62. (a) A.J. Mesiano, E.J. Beckman, A.J. Russell. Supercritical biocatalysis. *Chem. Rev.* 1999, 99, 623–633; (b) H.R. Hobbs, N.R. Thomas. Biocatalysis in supercritical fluids, in fluorous solvents, and under solvent-free conditions. *Chem. Rev.* 2007, 107, 2786–2820; (c) M. Habulin, M. Primozic, Z. Knez, Supercritical fluids as solvents for enzymatic reactions. *Acta Chim Slov.* 2007, 54, 667–677.
63. C.G. Laudani, M. Habulin, Z. Knez, G.D. Porta, E. Reverchon. Lipase-catalyzed long chain fatty ester synthesis in dense carbon dioxide: kinetics and thermodynamics. *J. Supercrit. Fluids* 2007, 41, 74–81.
64. M. Habulin, M. Primozic, Z. Knez. Enzymatic reactions in high-pressure membrane reactors. *Ind. Eng. Chem. Res.* 2005, 44, 9619–9625.
65. P. Lozano, A. B. Pérez-Marín, T. De Diego, D. Gómez, D. Paolucci-Jeanjean, M. P. Belleville, G.M. Rios, J.L. Iborra. Active membranes coated with immobilized *Candida antarctica* lipase B: preparation and application for continuous butyl butyrate synthesis in organic media. *J. Membr. Sci.* 2002, 201, 55–64.
66. (a) S. Kamat, J. Barrera, E.J. Beckman, A.J. Russell. Biocatalytic synthesis of acrylates in organic solvents and supercritical fluids 1. Optimization of enzyme environment. *Biotechnol. Bioeng.* 1992, 40, 158–166; (b) M. Habulin, Z. Knez. Activity and stability of lipases from different sources in supercritical carbon dioxide and near-critical propane. *J. Chem. Technol. Biotechnol.* 2001, 76, 1260–1266; (c) A. Striolo, A. Favaro, N. Elvassore, A. Bertucco, V. Di Notto. Evidence of conformational changes for protein films exposed to high-pressure CO_2 by FT-IR spectroscopy. *J. Supercrit. Fluids* 2003, 27, 283–295.
67. P. Lozano, A. Avellaneda, R. Pascual, J.L. Iborra. Stability of immobilized alpha-chymotrypsin in supercritical carbon dioxide *Biotechnol. Lett.* 1996, 18, 1345–1350; A. Giessauf, W. Magor, D.J. Steinberger, R. Marr. A study of hydrolases stability in supercritical carbon dioxide (SC-CO_2). *Enzyme Microb. Technol.* 1999, 24, 577–583.

68. W. Keim. Multiphase catalysis and its potential in catalytic processes: the story of biphasic homogeneous catalysis. *Green Chem.* 2003, 5, 105–111; M. Araia, S.I. Fujitaa, M. Shirai. Multiphase catalytic reactions in/under dense phase CO_2 *J. Supercrit. Fluids* 2009, 47, 351–356.
69. (a) P. Lozano, T. De Diego, D. Carrie, M. Vaultier, J.L. Iborra. Continuous green biocatalytic processes using ionic liquids and supercritical carbon dioxide. *Chem. Commum.* 2002, 692–693; (b) M.T. Reetz, W. Wiesenhofer, G. Francio, W. Leitner. Biocatalysis in ionic liquids: batchwise and continuous flow processes using supercritical carbon dioxide as the mobile phase. *Chem. Commun.* 2002, 992–993.
70. P. Lozano, T. De Diego, D. Carrie, M. Vaultier, J.L. Iborra. Lipase catalysis in ionic liquids and supercritical carbon dioxide at 150°C. *Biotechnol. Prog.* 2003, 19, 380–382.
71. (a) P. Lozano, T. De Diego, S. Gmouh, M. Vaultier, J.L. Iborra. Criteria to design green enzymatic processes in ionic liquid/supercritical carbon dioxide systems. *Biotechnol. Prog.* 2004, 20, 661–669; (b) P. Lozano, T. De Diego, D. Carrie, M. Vaultier, J.L. Iborra. Synthesis of glycidyl esters catalyzed by lipases in ionic liquids and supercritical carbon dioxide. *J. Mol. Cat. A: Chem.* 2004, 214, 113–119.
72. M.T. Reetz, W. Wiesenhofer, G. Francio, W. Leitner. Continuous flow enzymatic kinetic resolution and enantiomer separation using ionic liquid/supercritical carbon dioxide media. *Adv. Synth. Catal.* 2003, 345, 1221–1228.
73. E. Garcia-Junceda (Ed). *Multi-Steps Enzyme Catalysis: Biotransformations and Chemoenzymatic Synthesis*. Wiley-VCH: Weinheim, 2008.
74. P. Lozano, T. De Diego, M. Larnicol, M. Vaultier, J.L. Iborra. Chemoenzypmatic dynamic kinetic resolution of *rac*-1-phenylethanol in ionic liquids and ionic liquids/supercritical carbon dioxide systems. *Biotechnol. Lett.* 2006, 28, 1559–1565.
75. P. Lozano, T. De Diego, C. Mira, K. Montague, M. Vaultier, J.L. Iborra. Long term continuous green chemoenzymatic dynamic kinetic resolution of rac-1-phenylethanol using ionic liquids and supercritical carbon dioxide. *Green Chem.* 2009, 11, 538–542.
76. P. Lozano, J.M. Bernal, G. Sánchez-Gómez, G. López-López, M. Vaultier. How to produce biodiesel easily using a green biocatalytic approach in sponge-like ionic liquids. *Energy Environ. Sci.* 2013, 6, 1328–1338.
77. S. Serra, C. Fuganti, E. Brenna. Biocatalytic preparation of natural flavours and fragrances. *Trends Biotechnol.* 2005, 23, 193–198.
78. H. Ghamgui, M. Karra-Chaabouni, S. Bezzine, N. Miled, Y. Gargouri. Production of isoamyl acetate with immobilized Staphylococcus simulans lipase in a solvent-free system. *Enzyme Microb. Technol.* 2006, 38, 788–794; A. Guvenc, N. Kapucu, H. Kapucu, O. Aydogan, U. Mehmetoglu. Enzymatic esterification of isoamyl alcohol obtained from fusel oil: optimization by response surface methodolgy. *Enzyme Microb. Technol.* 2007, 40, 778–785.
79. P. Lozano, J.M. Bernal, R. Piamtongkam, D. Fetzer, M. Vaultier. One-phase ionic liquid reaction medium for biocatalytic production of biodiesel *ChemSusChem.* 2010, 3, 1359–1363; (b) P. Lozano, J.M. Bernal, M. Vaultier. Towards continuous sustainable processes for enzymatic synthesis of biodiesel in hydrophobic ionic liquids/supercritical carbon dioxide biphasic systems. *Fuel* 2011, 90, 3461–3467; P. Lozano, T. de Diego, J. L. Iborra and M. Vaultier. Use of ionic liquids for implementing a process for the preparation of biodiesel. *EP 2189535A1*, 2010

chapter four

Deep eutectic solvents

Promising solvents and nonsolvent solutions for biocatalysis

Pablo Domínguez de María

Contents

4.1 Relevance of biocatalysis in nonconventional media

The use of enzymes and whole cells for synthetic purposes has become nowadays a mature and well-established playground for the production of many industrially sound products, ranging from low-added-value bulk commodities to highly valuable optically active compounds. To this end, different types of enzymes may catalyze a broad number of different reactions and use an ample number of nonnatural substrates. A crucial key point to explain the present (increasing) trend of using enzymes for industrial syntheses is the development of molecular biology strategies that are enabling the "on demand" production of biocatalysts—through cloning and overexpression of desired genes—as well as the genetic design and improvement of a given biocatalyst to tailor it to the desired *industrial* reaction conditions (Bornscheuer et al. 2012; Drauz et al. 2012; Faber 2008; Gamenara et al. 2012; Liese et al. 2006; Reetz 2011, 2012; Whittall and Sutton 2012). Recent examples on using genetically designed transaminases for the competitive production of pharmaceuticals (Savile et al. 2010), or the engineering of whole cells to deliver alkenes via fermentation

(Domínguez de María 2011), are further novel outstanding examples adding more value to this emerging field.

In most of the cases, the choice of enzymes as biocatalysts is driven by the excellent selectivity (enantio-, regio-, or chemo-) that biocatalysts typically display, thus being advantageous when compared to other chemocatalysts. In addition, due to the often mild reaction conditions applied for biocatalytic reactions (e.g., ambient temperature and pressure, setup of aqueous conditions, etc.), the use of enzymes in organic synthesis has typically been regarded as green chemistry. Yet, despite those mild reaction conditions normally applied for enzymes, in recent years it has been pointed out that using aqueous solutions might not completely fulfill the green standards, increasingly required nowadays, as the formation of significant amounts of wastewater may be expected during downstream unit operations. In addition, from a practical viewpoint, many organic substrates are hardly soluble in aqueous solutions, and therefore enzymatic applications with high substrate loadings might be hampered. Thus, organic (co)solvents need to be added to aqueous solutions to favor that solubility.

To overcome these issues—trying to combine the usefulness of enzymes with emphasis on more sustainable chemistry—in recent years a number of biocatalytic applications pinpointing these sustainable premises have appeared. For aqueous biocatalysis, a range of biomass-derived organic solvents have been put forth, e.g., 2-methyltetrahydrofuran (2-MeTHF), or other glycerol-based solvents (Pérez-Sánchez et al. 2013; Pace et al. 2012; Hernáiz et al. 2010). These compounds may be used either as water-miscible cosolvents or directly as a second phase, to apply biocatalytic biphasic systems replacing hazardous petroleum-based (co)solvents for others with higher biodegradability and with (expected) diminished ecological footprints. In parallel to those efforts in aqueous solutions, biocatalysis in nonaqueous media—the so-called nonconventional media—has been a matter of intense research for decades as well, starting from the seminal work of Klibanov (2001) with hydrolases. This area is presently experiencing a *renaissance*, with emerging strategies like the use of neat substrates—water-free conditions with high substrate loadings—for whole-cell oxidoreductases (Jakoblinnert et al. 2011), or the genetic design of transaminases able to efficiently catalyze reactions in nonaqueous media (Mutti and Kroutil 2012; Truppo et al. 2012), among different examples. In a broad sense, several advantages may be envisaged from those systems: less wastewater formation and more straightforward work-up for organic solvents, higher substrate loadings and therefore better productivities for industrial applications, and wider technical options for the combination of biocatalysis and organic compounds, as nonaqueous media are set up, and thus problems derived by the low solubility of substrates are significantly circumvented.

Furthermore, apart from these more or less *classic* biocatalyses in nonconventional media, the last decade has witnessed the use of other neoteric solvents for nonaqueous biocatalytic applications. These are typically gathered under the brand name of ionic liquids (ILs) (Welton 1999). The potential of these ILs has been critically examined, covering, among others, broad applications in organic synthesis and catalysis (Wasserscheid and Welton 2007), analytic chemistry and polymer dissolution (Domínguez de María and Martinsson 2009), or biocatalysis, either as solvents for reactions (Domínguez de María 2012; Fernández-Álvaro and Domínguez de María 2012; Domínguez de María and Maugeri 2011; Lozano 2010) or in other nonsolvent *smart* applications, such as enzyme immobilization, enzyme coating, creative work-up strategies, etc. (Lozano et al. 2012a; Domínguez de María 2008). Different chapters of this book have described in detail the application of such ionic liquids to enzymatic reactions and biotechnology.

Mostly based on properties like their low flammability and largely diminished vapor pressure—especially when compared to *classic* organic solvents—ILs have often been generally regarded as green solvents. Certainly, this perception has contributed significantly to enhance the interest for research and applications of ILs, as it appeared that the combination of green chemistry and novel innovative solutions might be *merely* reached by means of these neoteric solvents. Unfortunately, years later it was realized that not *all* ILs can fall into the green solvent category. In this respect, based on the *E*-factor for the synthesis of ILs (Deetlefs and Seddon 2010), as well as in ecological prospects (ecotoxicity) that they show when spoiled in the environment (Domínguez de María 2012), a more realistic perspective on the strengths and weaknesses of ILs has emerged. But interestingly, herein another important property of ILs can be used: their wide tunability, allowing the production of a broad number of solvents by combining different anions and cations. Thus, research has recently focused on keeping the promising properties of ILs, while at the same time providing better environmental prospects by designing ILs with more appropriate environmental prognoses, both in their syntheses and in their fate. In this line, efforts in the development of bio-based ionic liquids—entirely composed of renewable resources—have been made, being (natural) amino acid outstanding starting materials for those purposes (Fukaya et al. 2007). For biocatalysis, the trend is to move from the second generation of ionic liquids (e.g., imidazolium-based ones) to the third generation (e.g., bio-based ILs) and the emerging deep eutectic solvents (DESs) (Domínguez de María 2012; Domínguez de María and Maugeri 2011; Gorke et al. 2010). These novel solvents are expected to keep many of the properties that the second generation of ILs display, but with more promising ecological and economic prognoses to envisage their practical application in the coming years. The next sections of this chapter will provide an overview of the emerging use of DESs in biocatalysis.

4.2 From knowledge on ionic liquids to DESs

As stated above, one of the current trends on ionic liquids and biotransformations is the assessment of novel biomass-derived, enzyme-friendly ILs that may provide better ecological footprints while, at the same time, being able to keep the promising properties that neoteric solvents may bring. With an analogous focus, another emerging type of neoteric solvents is represented by the so-called deep eutectic solvents (DESs). Albeit eutectic mixtures have been known for decades (Rastogi and Bassi 1964; Gambino et al. 1987), it has not been until the work of Abbott et al. (2003, 2004)—a decade ago—when DESs started to become popular within the scientific community. Surely the work of Abbott et al. appeared right in the time in which ILs had gained *momentum*. Since the first discoveries on DESs, a broad number of applications for these novel solvents have been reported, being used in many different areas like metal deposition, catalysis, biocatalysis, separation technology, etc. (Ruß and König 2012; Zhang et al. 2012; Carriazo et al. 2012).

In essence, DESs are mixtures of halide salts with a hydrogen bond donor (HBD) in different molar ratios (depending on each DES, an optimum ratio among components is expected). The combination of the halide with the HBD leads to an interaction of the HBD with the salt structure, thus disrupting its crystalline pattern by diminishing the electrostatic forces, and therefore triggering a decrease in the freezing point of the eutectic mixture. In many of these cases, the eutectic sink is so pronounced that liquid mixtures are formed at room temperature. For instance, the *classic* first example of a DES was composed of choline chloride (m.p. 302°C) and urea (m.p. 135°C) at different molar ratios. Remarkably, both compounds are solids at room temperature—with high melting points—but the mixture of them leads to a clear liquid with a freezing point of 12°C (at an optimal proportion ChCl:urea 1:2 mol:mol). At other proportions, different freezing points are obtained (Scheme 4.1).

The discovery of the ChCl:Ur DES (Scheme 4.1) (Abbott et al. 2003, 2004) stimulated that many other combinations of halide salts and HBDs were assessed by several groups for the formation of novel DESs. Herein, another outstanding case is represented by mixing choline chloride and glycerol (Abbott et al. 2011). In this case, the inherent high viscosity of glycerol is drastically diminished upon addition of choline chloride, leading to an easy-to-handle, inexpensive, and environmentally friendly ionic solvent. In an analogous line, the assessment of other bio-based HBDs has been studied (Maugeri and Domínguez de María 2012). Biomass-derived compounds like xylitol or levulinic acid may be used to form other inexpensive and biodegradable DESs in combination with choline chloride as halide salt. Moreover, the use of chiral compounds as DES components may lead to the formation of chiral solvents in a straightforward manner.

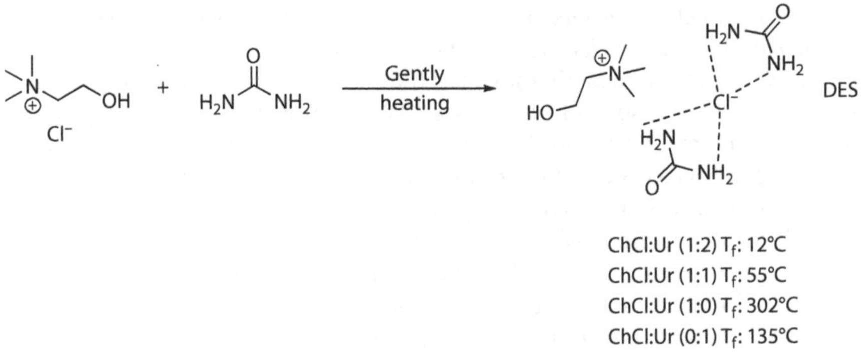

Scheme 4.1 Formation of DES ChCl:Urea (1:2) and postulated structure for the liquid DESs. (From Abbott, A. P. et al., *Chem. Commun.*, 70–71, 2003; Abbott, A. P. et al., *J. Am. Chem. Soc.*, 126, 9142–9147, 2004.)

As an example of that, isosorbide—another biomass-derived cyclic diol (Rose and Palkovits 2012)—may be used in combination with choline chloride as well to afford a chiral liquid at room temperature (Maugeri and Domínguez de María 2012). Since a significant amount of DES—the HBD part—is not ionic, there may be discrepancy about whether these solvents should be regarded as true ionic liquids. Far from these *academic* considerations, DESs may be defined as ionic solvents that actually share many of the properties of the other ILs—such as low flammability or volatility, as well as relatively high viscosities—while at the same time being often biodegradable and inexpensive.

As can be rapidly observed, another important asset for DESs is that they can be straightforwardly prepared by gently mixing the halide and HBD for several hours at moderate temperatures (up to 100°C), and without any purification steps, and therefore without waste formation. Furthermore, by looking at typical DES components, they are quite often formed by using inexpensive, readily available, and biodegradable structures. As an example, chlorine chloride—also termed vitamin B_4—is annually produced in huge quantities for animal feeding, and it is an essential human micronutrient as well (Blusztajn 1998). It must be noted, however, that most of the produced and commercialized choline chloride follows nowadays an ethylene-based petrochemical route, albeit future bio-based routes may be envisaged as well. Furthermore, preliminary studies on the toxicological effects of DESs have started to appear (Hayyan et al. 2013), providing promising features (yet still preliminary) on their toxicological effects toward different living organisms.

With regard to physicochemical properties, as a general trend DESs may dissolve components bearing hydrogen bond donors (e.g., carboxylic acids, sugars, alcohols, amines, etc.), whereas other non-hydrogen-bond donors (e.g., esters, ethers, ketones, etc.) typically form a clear second phase with DESs. This feature may be very useful for further downstream operation units, as extractive solvents like ethyl acetate, 2-MeTHF, acetone, or alkanes form a second phase (see also Section 4.4), and thus product extraction from DESs may be envisaged. Remarkably, in some specific cases, e.g., DESs formed with choline chloride and levulinic acid as HBD, the addition of an antisolvent (typically acetone) leads to the disruption of the ionic interaction, followed by the immediate quantitative precipitation and recovery of choline chloride (Maugeri and Domínguez de María 2012). On the other hand, all reported DESs are fully miscible with water, forming ionic solutions of the halide and HBD.

Following analogous considerations to other ionic liquids, it can be rapidly inferred that DESs may still keep a broad tunability of their properties, as halide salt and HBDs can be tailored and mixed in a broad number of combinations. Even the formation of ternary (or superior) systems might be envisaged, e.g., by combining two halide salts or several HBDs. Herein, starting from the first reported choline chloride-based DES (e.g., using urea or glycerol), a broad number of halide salts have been assessed in their capability to form DESs with different HBDs (Figure 4.1) (Ruß and König 2012; Zhang et al. 2012; Carriazo et al. 2012). In this respect, apart from choline chloride, other natural structures like (chiral) carnitine, betaine, or acetyl choline have also been successfully used to form DESs with different properties and viscosities. In addition to them, many other ammonium quaternary salts—with shorter or larger aliphatic chains—can also be employed, providing novel framework operations for different chemical problems. Likewise, several zinc, tin, or iron halides

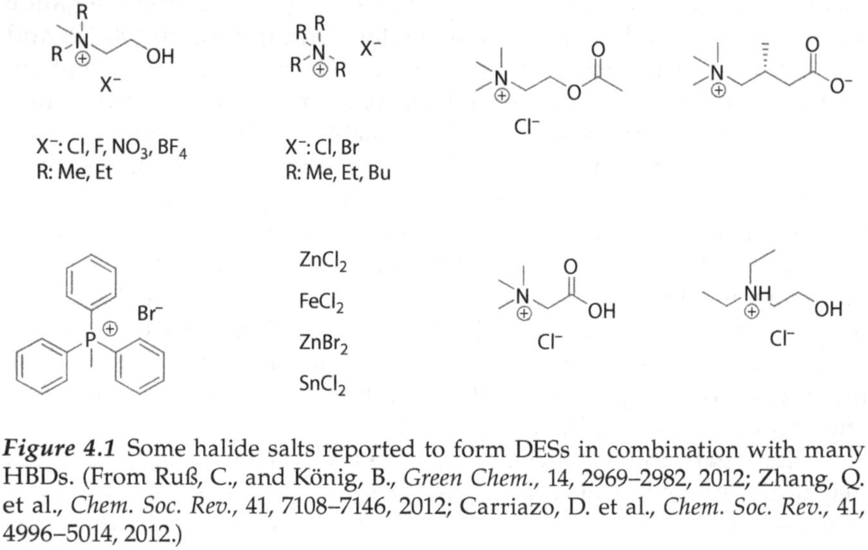

Figure 4.1 Some halide salts reported to form DESs in combination with many HBDs. (From Ruß, C., and König, B., *Green Chem.*, 14, 2969–2982, 2012; Zhang, Q. et al., *Chem. Soc. Rev.*, 41, 7108–7146, 2012; Carriazo, D. et al., *Chem. Soc. Rev.*, 41, 4996–5014, 2012.)

may lead to DES formation, eventually enabling the formation of solvent-catalyst mixtures with novel properties.

Likewise, the above-reported versatility to choose halide salts for DES formation can be used for the choice of hydrogen bond donor structures. In this area, many alcohols, polyols, sugars, carboxylic acids, and amines and amides have been successfully employed (Figure 4.2) in different proportions and combined with several halide salts (Ruß and König 2012; Zhang et al. 2012; Carriazo et al. 2012). As stated above, the introduction of chiral structures (e.g., isosorbide) provides a simple and elegant entry for the formation of chiral solvents. In this respect, it is tempting to assess many biomass-derived structures—typically bearing alcohols, carboxylic acids, etc.—to create a new toolbox of tunable and biodegradable ionic solvents easily accessible at large scale. Furthermore, the area of DESs has been expanded to other amino acids—carboxylic acid mixtures as well—with promising applications for biorefineries and lignocellulose processing (Francisco et al. 2012).

As stated above, the first DESs were reported a decade ago (Abbott et al. 2003, 2004). Despite this relative novelty, an impressive number of applications of several DESs in different areas of chemistry have already been reported, covering metal deposition, their use as solvents for organic synthesis and catalysis, as well as for separation units (Ruß and König 2012; Zhang et al. 2012; Carriazo et al. 2012). Furthermore, an interesting area of research is the extraction of natural products, due to the capacity

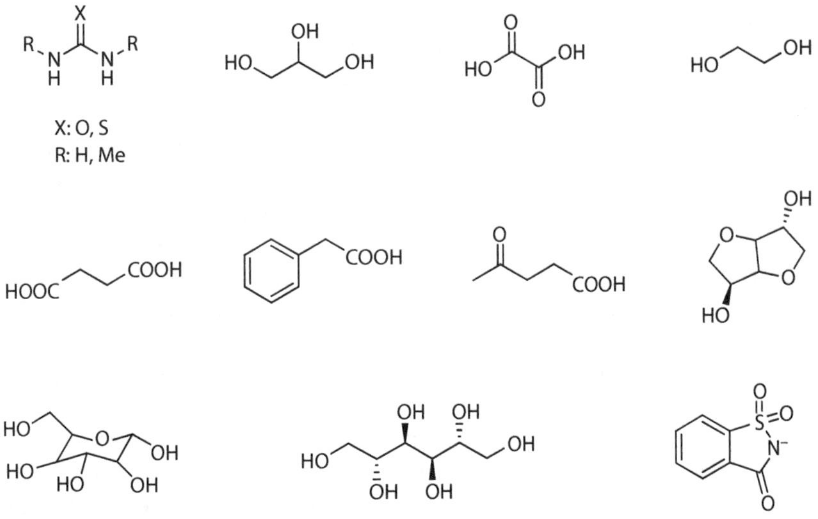

Figure 4.2 Several selected structures used as HBDs to form DESs. (From Ruß, C., and König, B., *Green Chem.*, 14, 2969–2982, 2012; Zhang, Q. et al., *Chem. Soc. Rev.*, 41, 7108–7146, 2012; Carriazo, D. et al., *Chem. Soc. Rev.*, 41, 4996–5014, 2012.)

of many DESs to dissolve natural products, as most of these compounds bear hydrogen bond donor molecules (Hertel et al. 2012; van Spronsen et al. 2011). Similarly, the possibility of dissolving lignin with some DESs, whereas other main parts of lignocellulose—polysaccharides, mainly hemicellulose and cellulose—remain suspended on it (Francisco et al. 2012), may deliver new innovative entries for biorefineries, using biomass-derived neoteric solvents. Promising delignification processes of agricultural residues have been suggested by means of these DESs (Francisco et al. 2012). Actually, such capacity for dissolving many water-immiscible natural products has led to the hypothesis that natural reactions would actually take place within DESs that are in situ formed within living cells (e.g., by combining natural ammonium salts and sugars in microenvironments) (Choi et al. 2011). DESs have also been assessed in several biocatalytic reactions as well—aiming at combining green catalysts with green solvents—as will be discussed in depth in the next sections.

4.3 DESs in biocatalysis: Solvents and cosolvents for enzyme-catalyzed reactions

As reported above, and in many other chapters of this book, the use of ionic liquids in biotransformations has led to many innovative applications, combining the broad tunability of ILs with practical options in biocatalysis (solvents, immobilizing supports, ILs for coatings, ILs for extractions, etc.). In the quest for novel neoteric solvents that could provide economic and environmentally friendly prognoses, several research groups have started to assess different DESs in biocatalysis as well. Remarkably, at first sight it might appear that the use of these solvents for enzyme-based processes could not deliver any beneficial effect. Actually, high concentrations of choline chloride are deleterious for enzymes. Likewise, urea is a well-known and widely used denaturizing agent for proteins. On the other hand, there are some examples of protic ionic liquids able to stabilize enzymes (Attri et al. 2011; Falcioni et al. 2010), when one would expect denaturizing effects. Therefore, it was not obvious that DESs might represent a useful reaction media for biocatalysis in nonconventional media.

Interestingly, several studies have confirmed that many enzymes are actually very stable and active in different DESs, thus opening new options for biocatalysis in those emerging nonconventional media, making use of biomass-derived, tunable, and inexpensive solvents. In this area, the first proof of concept was given by the Kazlauskas group studying different lipases for several synthetic reactions—transesterifications, aminolysis, and lipase-mediated epoxidations—in different DESs and comparing them to toluene as a prototypical nonaqueous solvent for these enzymes (Gorke et al. 2008, 2009). A comparison of transesterifications

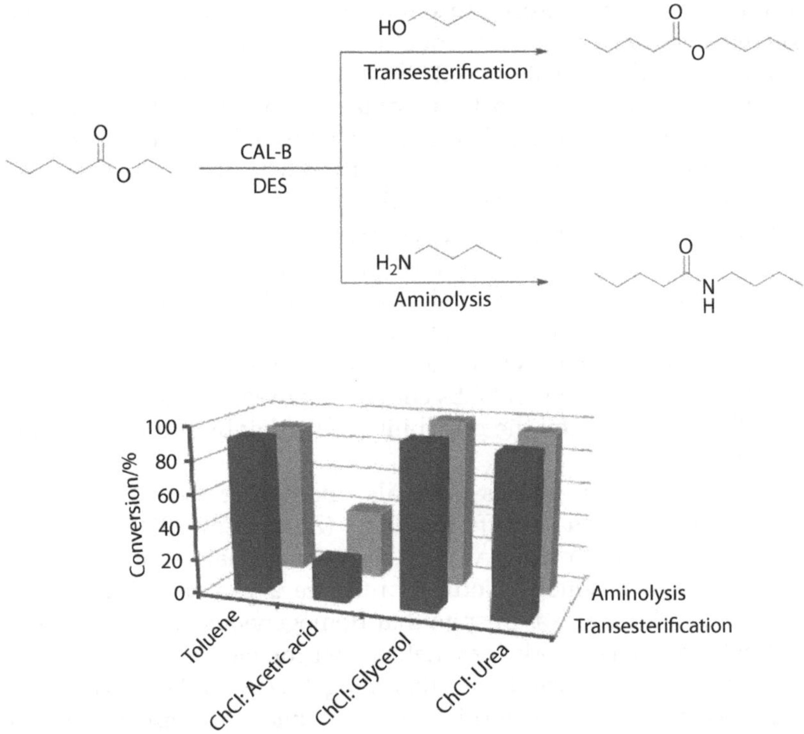

Figure 4.3 Assessment of different choline chloride-based DESs in lipase-catalyzed transesterifications and aminolysis. (From Gorke, J. et al., *Chem. Commun.*, 1235–1237, 2008; Gorke, J. et al., Enzymatic Processing in Deep Eutectic Solvents, US2009/0117628, 2009.)

and aminolysis in different choline chloride-based DESs using immobilized lipase from *Candida antarctica* was conducted (Figure 4.3).

As observed, immobilized *C. antarctica* lipase B (CAL-B) displayed excellent activities in DESs containing choline chloride and glycerol or urea as HBDs, leading to full conversions, and analogous to results observed for toluene. In other, more acidic DESs (using acetic acid as HBD), lower conversions were reached, presumably due to the high acidity of the reaction media. Thus, it could be confirmed that DESs containing denaturizing agents for proteins were actually valid environments for enzyme catalysis (at least for lipases). Moreover, other lipases like lipase A from *C. antarctica* or *Pseudomonas cepacia* lipase also displayed activities in different DESs. Furthermore, other DESs containing ethylammonium chloride as halide salt displayed positive results as well (Gorke et al. 2008, 2009). Another important point for consideration is whether the used HBDs—alcohols, carboxylic acids—could also play the role of substrates for the lipases, and thus interfere with the desired reaction. Herein, secondary (trans)esterification reactions have been found for ethylene glycol and glycerol, the latter in lower extent (Gorke et al. 2008, 2009). In any case, the transesterification rates for butanol as actual substrate, compared to those of HBDs (ethylene glycol or glycerol), were much higher in DESs—that is, displaying better selectivities for the desired alcohol—than when both substrates were added in toluene as reaction media. Other results reported recently by the Villeneuve group (Durand et al. 2012) are consistent with these observations. In CAL-B-catalyzed transesterifications using different vinyl esters as acyl donors, as well as several HBDs—oxalic acid, malonic acid, or ethylene glycol—different esters of these compounds were observed and chemically characterized. A potential solution for that may be the setup of lipase-catalyzed reactions in DESs containing urea, instead of alcohols or carboxylic acids, whereby the possibility of further by-product formation is excluded.

In the same area, lipases have also been assessed as biocatalysts for biodiesel formation in different DESs (Zhao et al. 2011a, 2013; Zhao and Baker 2013). When using wastes as raw materials—e.g., cooking oil and agricultural residues—biodiesel may be a promising tool for the valorization of local areas. Despite that homogeneous chemical catalysts (e.g., NaOH or KOH) work reasonably well for biodiesel synthesis, the waste production associated with them, together with the need of implementing prestep units for esterification reactions—to eliminate the excess of free fatty acids (FFAs) typically present in oil wastes, gives rise to the quest of novel alternative catalysts for biodiesel synthesis, which might catalyze both esterifications and transesterifications at the same time. In this regard, lipases may be excellent options, efficient and associated with high environmental standards, provided that adequate enzyme immobilization strategies are set. Such immobilization may enable (expensive) biocatalysts to be reused for a number of cycles, considering the expected low price that biodiesel—as fuel—must have. Herein, the Zhao group has studied different DESs for such purposes, reporting the use of choline acetate chloride and glycerol as a mixture with a diminished viscosity, thus being a promising alternative for biodiesel synthesis. In that mixture, immobilized CAL-B remained stable and active for several cycles (Zhao et al. 2011a, 2013; Zhao and Baker 2013). Even real substrates like soybean oil could be satisfactorily used for such purposes. The use of DESs in the biodiesel area has attracted interest in separations as well, as will be reported in the next section.

Apart from hydrolases, proteases have been used in DESs as well. In this area, it is worth mentioning some research that was performed during the 1990s—before DESs were described—in which the direct mixture of substrates (amino acids and peptides) led to the formation of eutectic derivatives whereby protease-catalyzed peptide synthesis can be performed properly (López-Fandino et al. 1994a, 1994b; Gill and Vulfson 1993, 1994; Jorba et al. 1995; Gill and Valivety 2002). The approach is highly efficient, and it is surprising the scarce interest that this concept has attracted for practical biocatalysis. In a nutshell, amino acids and peptides can form eutectics and analogous solid-liquid complexes when they are mixed in different molar ratios. In some occasions the addition of other "adjuvants"—typically water or ethanol, up to 20% w/w—may contribute to the formation of the eutectics. Especially when hydrophobic and hydrophilic substrates are concomitantly used, the eutectic formation may contribute to solubilize both to create a highly useful reaction media. By means of this approach, several highly valuable oligopeptides were synthesized using α-chymotrypsin, chymopapain, subtilisin, papain, or thermolysin, thus showing that the concept may be actually widely applicable for many enzymes. As a clear advantage of those systems, high substrate loadings may be *liquefied*, thus enhancing the productivity that the biocatalytic system may display. As another asset of these strategies, no organic solvents are used—except water or ethanol as adjuvants—and therefore beneficial *E*-factors may be expected from the application of these technologies. Some of these approaches were assessed at the pilot-scale level, and under optimized conditions, impressive productivities were achieved in each of the biocatalytic steps (Scheme 4.2) (Gill and Valivety 2002).

More recently, the Zhao group has assessed the use of subtilisin and α-chymotrypsin in different DESs using choline salts—chloride or acetate form—and glycerol as HBD. By properly adding amounts of water, high transesterification activities were observed, with high selectivities (>99%). Enzymes were immobilized on chitosan as support, and under these conditions promising reaction media for biocatalysis could be shown (Zhao et al. 2011b). Taken together, it appears that proteases may find very promising novel media by using DESs and other eutectic mixtures, even for their consideration at the industrial level as well.

As stated above, all reported DESs are fully miscible with aqueous solutions. Therefore, it is questionable whether aqueous solutions of DESs may still be regarded as ionic solvents, rather than as aqueous ionic solutions. Nevertheless, there are some examples in the literature in which the study of biocatalysis in aqueous solutions containing DESs may be of practical interest. For instance, epoxide hydrolases display higher regioselectivities in such solutions, yet at the cost of lower activities and affinities of the enzyme toward substrates. Interestingly, the addition of DESs as cosolvents (in this case, choline chloride combined with glycerol, urea, or ethylene glycol) led to a higher solubility of the substrate under these aqueous solutions (Lindberg et al. 2010).

Likewise, cellulases have been genetically engineered to display higher activities in polysaccharide depolymerization in aqueous solutions containing rests of DESs, or in salty effluents like seawater (Lehmann

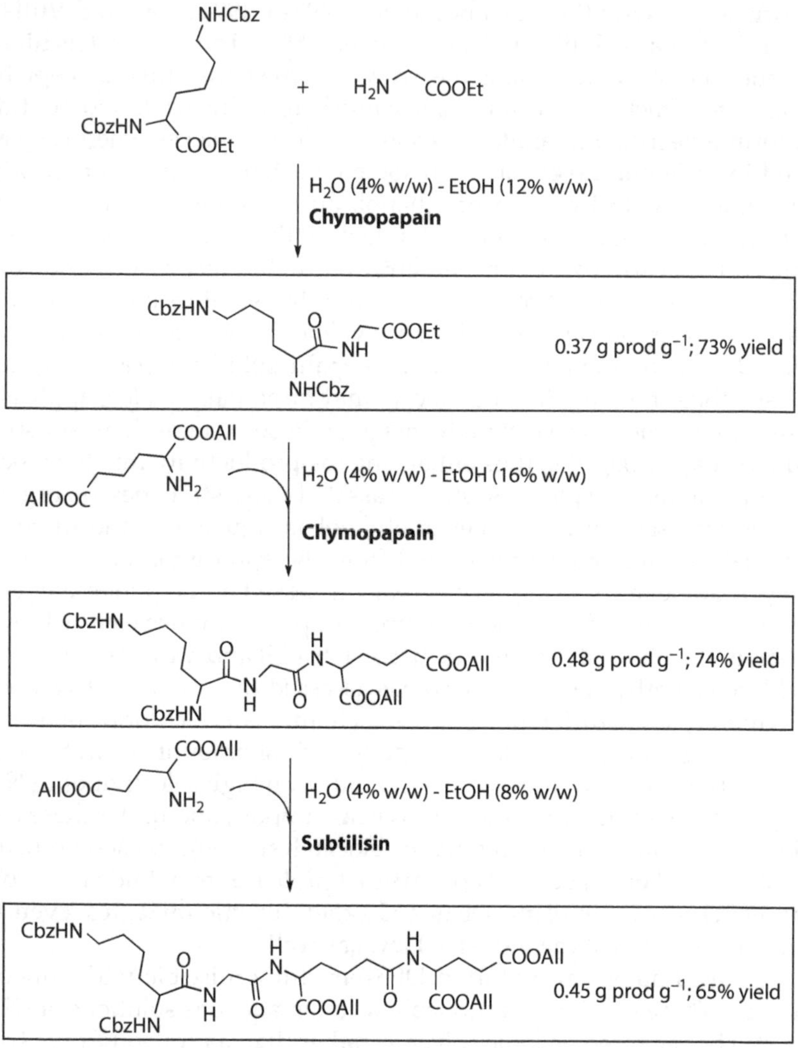

Scheme 4.2 Synthesis of an oligopeptide using several proteases and forming eutectic mixtures with substrates. (From Gill, I., and Valivety, R., *Org. Proc. Res. Dev.*, 6, 684–691, 2002.)

et al. 2012). The provision of cheap fermentable sugars from lignocellulose represents one of the current bottlenecks in biorefineries. Herein, the discovery that certain ionic liquids may dissolve different parts of the lignocellulose triggered research to combine cellulases and ionic liquids, with several outstanding examples reported in the open literature (Lozano et al. 2011, 2012b). Following a similar rationale, several DESs have been reported as solubilizers of parts of lignocellulose (Francisco et al. 2012).

Upon dissolution of biomass in ILs, the addition of an antisolvent—aqueous solutions—precipitates polysaccharides in an amorphous form, being then prone to be efficiently depolymerized by cellulases. Thus, envisaging biorefineries, it may be expected that aqueous effluents containing rests of either ionic liquids or DESs may be the common media for glycosidic enzymes in the future. Therefore, aiming at training cellulases under these *real* and challenging conditions, aqueous solutions of water containing choline chloride and glycerol (up to 20%) were used for directed evolution programs. Several variants of the enzyme showed higher activities and stabilities under these conditions (Lehmann et al. 2012), thus demonstrating that it is actually possible to generate enzyme variants displaying powerful competences in these lines.

Finally, another important part of biocatalysis is the setup of whole-cell-based systems. Notably, it is possible to perform enzymatic enantioselective reductions using whole cells in (water-free) neat substrate conditions, showing that cells are actually stable and enzymes active (Jakoblinnert et al. 2011). Albeit there are no current examples on using DESs and whole cells for biotransformations, a study showed that whole cells actually keep their integrity when lyophilized in DES media (Gutiérrez et al. 2010). This fact might be the basis for future whole-cell-based biocatalysis in nonconventional media using DESs.

4.4 Nonsolvent applications of DESs in biocatalysis: DESs as separating agents

Apart from their use as *mere* solvents, ionic liquids have found many other nonsolvent applications in biotransformations, enabling enzyme immobilization or coating, substrate engineering, or an ample number of creative membrane-based and separation technologies (Lozano et al. 2012a; Domínguez de María 2008, 2012). In an analogous line, DESs—as ionic solvents—may be envisaged as *something more than solvents,* and thus further *smarter* uses can be foreseen for these emerging mixtures.

In this regard, several applications for separation technologies may be based on the solubilizing capacity that DESs may have toward different chemicals. As stated above, DESs can dissolve hydrogen bond donor compounds, such as alcohols, amines, carboxylic acids, etc. Conversely, non-hydrogen-bond donor structures, such as esters, aldehydes, ketones, etc., tend to form a second phase with DESs. Focusing on this property, simple choline chloride–glycerol DESs (and others) have been used to remove the remnant of glycerol during biodiesel synthesis. Once biodiesel is synthetized from natural oils and methanol or ethanol—using chemocatalysts or lipases—glycerol is formed as the by-product, and it appears mixed with the biodiesel (the fatty acid methyl esters (FAMEs)). Upon centrifugation,

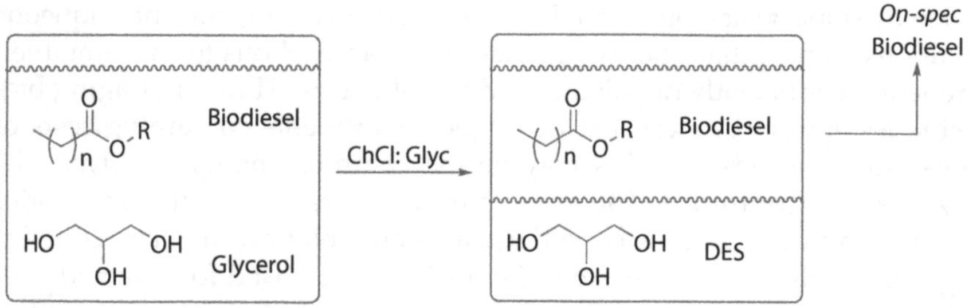

Scheme 4.3 Conceptual approach for the elimination of glycerol from biodiesel mixtures using DESs. (See Abbott, A. P. et al., 2007; Hayyan, M. et al., 2010; Shahbaz, K. et al., 2010, 2011a, 2011b).

a significant part of that glycerol and remaining alcohols (ethanol or methanol) can be removed from the biodiesel. However, glycerol damages the engine by impeding the high-pressure injection system. Conclusively, biodiesel regulations (EN 14214 and ASTM D6751) oblige that more glycerol must be removed from the final fuel composition. Apart from other approaches, it was realized that DESs do not dissolve biodiesel—as they are fatty acid esters without hydrogen bond donor parts—but do extract glycerol and other alcohols from the biodiesel mixture. With this approach, a DES treatment leads to *on-spec* conditions for the biofuel (Abbott et al. 2007; Hayyan et al. 2010; Shahbaz et al. 2010, 2011a, 2011b) (Scheme 4.3). An analogous DES-based approach for extraction-separation has been reported for phenol derivatives dissolved in mineral oils (Pang et al. 2012).

In the same area of separations, very recently other applications combining enzymes and DESs as nonsolvent agents have been reported by the Domínguez de María group. One of them focuses on the solvent-free lipase-catalyzed kinetic resolution of racemic alcohols, using vinyl esters as acyl donors (Maugeri et al. 2012). The setup of a solvent-free system allows high substrate loadings (up to ca. 300 g L^{-1} racemate), and under optimized conditions, immobilized lipase B from *C. antarctica* efficiently catalyzes the reaction, leading to excellent enantioselectivities (>99%) and quantitative conversions for a kinetic resolution (50%). Therefore, a mixture of an enantiomer as alcohol, and another one as ester, is achieved at the end of the reaction. To separate these mixtures, chromatographic techniques are typically applied. Herein, by making use of choline chloride–glycerol DESs, kinetically resolved esters form a second phase—due to the absence of the hydrogen bond donor part—whereas the enantiopure alcohol is dissolved in DESs (Scheme 4.4). After several extractive cycles, high purities on esters (up to 99%) with high efficiencies (80–90% total ester extracted) were achieved under still nonoptimized conditions (Maugeri et al. 2012).

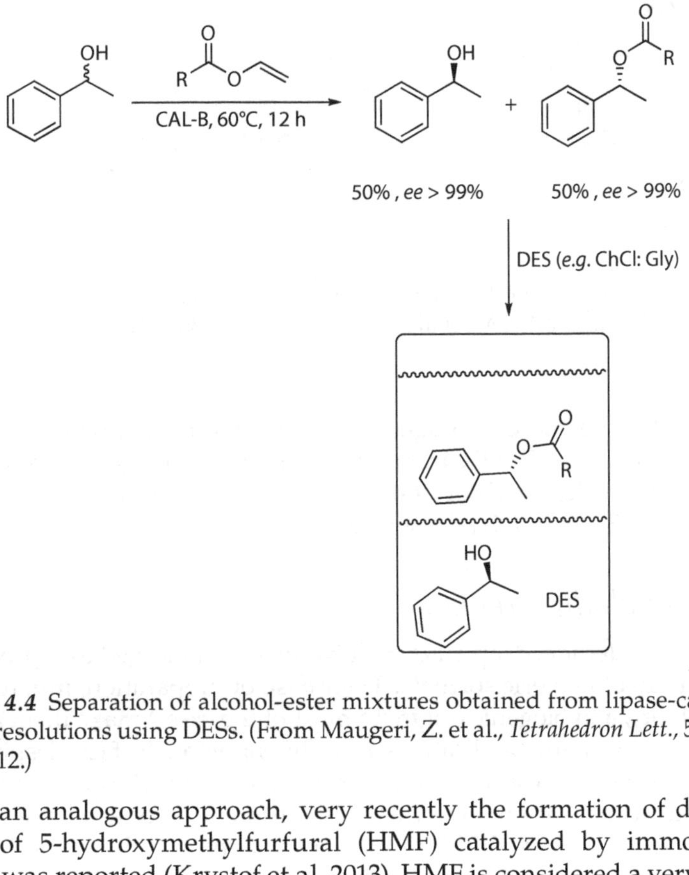

Scheme 4.4 Separation of alcohol-ester mixtures obtained from lipase-catalyzed kinetic resolutions using DESs. (From Maugeri, Z. et al., *Tetrahedron Lett.*, 53, 6968–6971, 2012.)

In an analogous approach, very recently the formation of different esters of 5-hydroxymethylfurfural (HMF) catalyzed by immobilized CAL-B was reported (Krystof et al. 2013). HMF is considered a very promising biomass-derived platform chemical, produced form the triple dehydration of hexoses under acidic conditions (van Putten et al. 2013; Grande et al. 2012). Despite there being presently no industrial process to afford HMF from biomass, it is expected that this situation will change in the next years. Notably, HMF can be derivatized to deliver a broad array of useful chemicals, ranging from fuels additives to monomers for fibers, surfactants, etc. In this respect, the provision of different HMF esters may be an interesting field for its valorization. Yet, due to the inherent reactivity of HMF, process conditions must be mild, to avoid the formation of by-products, oligomerizations, etc., that may hamper a competitive process. Very recently, lipases have been reported as very useful biocatalysts for the HMF esterification in solvent-free conditions, using different acyl donors and with high substrate loadings and productivities (Krystof et al. 2013). Herein, typically yields of esters of ca. 90% are achieved, whereby a remnant of HMF is typically observed. To separate that mixture, DESs

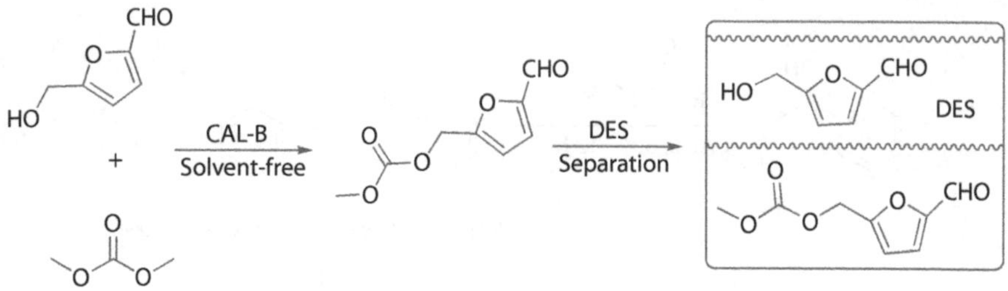

Scheme 4.5 Synthesis of HMF esters catalyzed by CAL-B, and separation of esters using DES. (From Krystof, M. et al., *ChemSusChem*, 6, 630–634, 2013.)

turned out to be successful agents. Using dimethylcarbonate (DMC) as an acyl donor, pure HMF-carbonate esters were achieved with a simple extraction with choline chloride-based DESs (using glycerol, urea, or xylitol as HBDs) (Scheme 4.5) (Krystof et al. 2013).

4.5 *Concluding remarks*

Over the last decade, deep eutectic solvents have emerged as a promising novel family of neoteric solvents. Their ease of preparation, inherent biodegradability, and biogenic origin (in most of the cases) make them appealing options for many applications in different areas. In fact, despite their novelty, several interesting proofs of principle using DESs have appeared in the literature. For enzyme catalysis, different enzymes have already shown activities and stabilities in these solvents, and therefore their use as nonconventional media for synthetic purposes seems to be a realistic option. Lipases, proteases, and epoxide hydrolases have been assessed. In addition, whole cells seem to keep their integrity when lyophilized in these solvents. As an asset for their use, the capability of DESs to dissolve many organic compounds—which are typically not soluble in many other media—opens options for further valorization and practical applicability. Furthermore, apart from being used as reaction media, DESs have also shown some promising nonsolvent applications. Thus, the possibility of separating hydrogen bond donor molecules from nondonor ones allows straightforward chemical separations, as illustrated for glycerol and biodiesel, for enzyme-catalyzed kinetic resolutions, as well as for esters of 5-hydroxymethylfurfural. Being an emerging area, it may be expected that novel applications and DESs will be disclosed in the coming years. Together with the third generation of ionic liquids—greener, economic, and versatile—the field of neoteric solvents and biotransformations continues its growth in the quest of holistic practical applications in sustainable chemistry.

References

Abbott, A.P., Boothby, D., Capper, G., Davies, D.L., and Rasheed, R.K. 2004. Deep eutectic solvents formed between choline chloride and carboxylic acids: versatile alternatives to ionic liquids. *J. Am. Chem. Soc.* 126: 9142–9147.

Abbott, A.P., Capper, G., Davies, D.L., Rasheed, R.K., and Tambyrajah, V. 2003. Novel solvent properties of choline chloride/urea mixtures. *Chem. Commun.* 70–71.

Abbott, A.P., Cullis, P.M., Gibson, M.J., Harris, R.C., and Raven, E. 2007. Extraction of glycerol from biodiesel into a eutectic based ionic liquid. *Green Chem.* 9: 868–872.

Abbott, A.P., Harris, R.C., Ryder, K.S., D'Agostino, C., Gladden, L.F., and Mantle, M.D. 2011. Glycerol eutectics as sustainable solvent systems. *Green Chem.* 13: 82–90.

Attri, P., Venkatesu, P., Kumar, A., and Byrne, N. 2011. A protic ionic liquid attenuates the deleterious actions of urea on α-chymotrypsin. *Phys. Chem. Chem. Phys.* 13: 17023–17026.

Blusztajn, J.K. 1998. Choline, a vital amine. *Science* 281: 794–795.

Bornscheuer, U.T., Huisman, G., Kazlauskas, R.J., Lutz, S., Moore, J., and Robins, K. 2012. Engineering the third wave in biocatalysis. *Nature* 485: 185–194.

Carriazo, D., Serrano, M.C., Gutiérrez, M.C., Ferrer, M.L., and del Monte, F. 2012. Deep-eutectic solvents playing multiple roles in the synthesis of polymers and related materials. *Chem. Soc. Rev.* 41: 4996–5014.

Choi, Y.H., van Spronsen, J., Dai, Y., Verberne, M., Hollmann, F., Arends, I.W.C.E., Witkamp, G.J., and Verpoorte, R. 2011. Are natural deep eutectic solvents the missing link in understanding cellular metabolism and physiology? *Plant. Physiol.* 156: 1701–1705.

Deetlefs, M., and Seddon, K.R. 2010. Assessing the greenness of some typical laboratory ionic liquids preparations. *Green Chem.* 12: 17–30.

Domínguez de María, P. 2008. Non-solvent applications of ionic liquids in biontransformations and organocatalysis. *Angew. Chem. Int. Ed.* 47: 6960–6968.

Domínguez de María, P. 2011. Recent developments in the biotechnological production of hydrocarbons: paving the way for bio-based platform chemicals. *ChemSusChem* 4: 327–329.

Domínguez de María, P., ed. 2012. *Ionic liquids in biotransformations and organocatalysis: solvents and beyond.* John Wiley & Sons, Hoboken, NJ.

Domínguez de María, P., and Martinsson, A. 2009. Ionic-liquid-based method to determine the degree of esterification in cellulose fibers. *Analyst* 134: 493–496.

Domínguez de María, P., and Maugeri, Z. 2011. Ionic liquids in biotransformations: from proof-of-concept to emerging deep-eutectic-solvents. *Curr. Opin. Chem. Biol.* 15: 220–225.

Drauz, K.H., Gröger, H., and May, O. 2012. *Enzyme catalysis in organic synthesis.* Wiley-VCH, Weinheim.

Durand, E., Lecomte, J., Baréa, B., Piombo, G., Dubreucq, E., and Villeneuve, P. 2012. Evaluation of deep eutectic solvents as new media for *Candida antarctica* B lipase catalyzed reactions. *Process. Biochem.* 47: 2081–2089.

Faber, K. 2008. *Biotransformations in organic synthesis.* Springer, Berlin.

Falcioni, F., Housden, H.R., Ling, Z., Shimizu, S., Walker, A.J. and Bruce, N.C. 2010. Soluble, folded and active subtilisin in a protic ionic liquid. *Chem. Commun.* 46: 749–751.

Fernández-Álvaro, E., and Domínguez de María, P. 2012. Ionic liquids in biocatalytic oxidations: from non-conventional media to non-solvent applications. *Curr. Org. Chem.* 16: 2492–2507.

Francisco, M., van den Bruinhorst, A., and Kroon, M.C. 2012. New natural and renewable low transition temperature mixtures (LTTMs): screening as solvents for lignocellulose biomass processing. *Green Chem.* 14: 2153–2157.

Fukaya, Y., Lizuka, Y., Sekihawa, K., and Ohno, H. 2007. Bio ionic liquids: room temperature ionic liquids composed wholly of biomaterials. *Green Chem.* 9: 1155–1157.

Gambino, M., Gaune, P., Nabavian, M., Gaune-Escard, M., and Bros, J.P. 1987. Enthalpie de fusion de l'uree et de quelques mélanges eutectiques a base d'uree. *Thermochim. Acta* 111: 37–47.

Gamenara, D., Seoane, G., Saenz Méndez, P., and Domínguez de María, P. 2012. *Redox biocatalysis: fundamentals and applications*. John Wiley & Sons, Hoboken, NJ.

Gill, I., and Valivety, R. 2002. Pilot-scale enzymatic synthesis of bioactive oligopeptides in eutectic-based media. *Org. Proc. Res. Dev.* 6:684–691.

Gill, I., and Vulfson, E.N. 1993. Enzymatic synthesis of short peptides in heterogeneous mixtures of substrates. *J. Am. Chem. Soc.* 115: 3348–3349.

Gill, I., and Vulfson, E. 1994. Enzymic catalysis in heterogeneous eutectic mixtures of substrates. *Trends Biotechnol.* 12: 118–122.

Gorke, J.T., Kazlauskas, R.J., and Srienc, F. 2009. Enzymatic processing in deep eutectic solvents. US2009/0117628.

Gorke, J.T., Srienc, F., and Kazlauskas, R.J. 2008. Hydrolase-catalyzed biotransformations in deep eutectic solvents. *Chem. Commun.* 1235–1237.

Gorke, J., Srienc, F., and Kazlauskas, R.J. 2010. Toward advanced ionic liquids. Polar enzyme-friendly solvents for biocatalysis. *Biotechnol. Bioprocess. Eng.* 15: 40–53.

Grande, P.M., Bergs, C., and Domínguez de María, P. 2012. Chemo-enzymatic conversion of glucose into 5-hydroxymethylfurfural in seawater. *ChemSusChem* 5: 1203–1206.

Gutiérrez, M.C., Ferrer, M.L., Yuste, L., Rojo, F., and del Monte, F. 2010. Bacteria incorporation in deep eutectic solvents through freeze-drying. *Angew. Chem. Int. Ed.* 49: 2158–2162.

Hayyan, M., Hashim, M.A., Hayvan, A., Al-Saadi, M.A., AlNashef, I.M., Mirghani, M.E.S., and Saheed, O.K. 2013. Are deep eutectic solvents benign or toxic? *Chemosphere* 90: 2193–2195.

Hayyan, M., Mjalli, F.S., Ali Hashim, M., and AlNashef, I.M. 2010. A novel technique for separating glycerine from palm oil-based biodiesel using ionic liquids. *Fuel Proc. Technol.* 91: 116–120.

Hernáiz, M.J., Alcántara, A.R., García, J.I., and Sinisterra, J.V. 2010. Applied biotransformations in green solvents. *Chem. Eur. J.* 16: 9422–9437.

Hertel, R.M., Bommarius, A.S., Realff, M.J., and Kang, Y. 2012. Deep eutectic solvent systems and methods. WO2012/145522.

Jakoblinnert, A., Mladenov, R., Paul, A., Sibilla, F., Schwaneberg, U., Ansorge-Schumacher, M.B., and Domínguez de María, P. 2011. Asymmetric reduction of ketones with recombinant *E. coli* whole cells in neat substrates. *Chem. Commun.* 47: 12230–12232.

Jorba, X., Gill, I., and Vulfson, E.N. 1995. Enzymatic synthesis of the delicious peptide fragments in eutectic mixtures. *J. Agric. Food Chem.* 43: 2536–2541.

Klibanov, A.M. 2001. Improving enzymes by using them in organic solvents. *Nature* 409: 241–246.

Krystof, M., Pérez-Sánchez, M., and Domínguez de María, P. 2013. Lipase-catalyzed (trans)esterification of 5-hydroxymethylfurfural and separation of HMF-esters using deep-eutectic-solvents. *ChemSusChem*. 6: 630–634.

Lehmann, C., Sibilla, F., Maugeri, Z., Streit, W.R., Domínguez de María, P., Martinez, R., and Schwaneberg, U. 2012. Reengineering CelA2 cellulase for hydrolysis in aqueous solutions of deep eutectic solvents and concentrated seawater. *Green Chem.* 14: 2719–2726.

Liese, A., Seelbach, K., and Wandrey, C. 2006. *Industrial biotransformations*. 2nd ed. Wiley-VCH, Weinheim.

Lindberg, D., de la Fuente Revenga, M., and Widersten, M. 2010. Deep eutectic solvents are viable cosolvents for enzyme-catalyzed epoxide hydrolysis. *J. Biotechnol.* 147: 169–171.

López-Fandino, R., Gill, I., and Vulfson, E.N. 1994a. Enzymatic catalysis in heterogeneous mixtures of substrates: the role of the liquid phase and the effects of "adjuvants." *Biotechnol. Bioeng.* 43: 1016–1023.

López-Fandino, R., Gill, I., and Vulfson, E.N. 1994b. Protease-catalyzed synthesis of oligopeptides in heterogeneous substrate mixtures. *Biotechnol. Bioeng.* 43: 1024–1030.

Lozano, P. 2010. Enzymes in neoteric solvents: from one-phase to multiphase systems. *Green Chem.* 12: 555–569.

Lozano, P., Bernal, B., Bernal, J.M., Pucheault, M., and Vaultier, M. 2011. Stabilizing immobilized cellulase by ionic liquids for saccharification of cellulose solutions in 1-butyl-3-methylimidazolium chloride. *Green Chem.* 13: 1406–1410.

Lozano, P., Bernal, J.M., and Navarro, A. 2012a. A clean enzymatic process for producing flavor esters by direct esterification in switchable ionic liquid/solid phases. *Green Chem.* 14: 3026–3033.

Lozano, P., Bernal, B., Recio, I., and Belleville, M.P. 2012b. A cyclic process for full enzymatic saccharification of pretreated cellulose with full recovery and reuse of the ionic liquid 1-butyl-3-methylimidazolium chloride. *Green Chem.* 14: 2631–2637.

Maugeri, Z., and Domínguez de María, P. 2012. Novel choline-chloride-based deep eutectic solvents with renewable hydrogen bond donors: levulinic acid and sugar-based polyols. *RSC Adv.* 2: 421–425.

Maugeri, Z., Leitner, W., and Domínguez de María, P. 2012. Practical separation of alcohol-ester mixtures using deep eutectic solvents. *Tetrahedron Lett.* 53: 6968–6971.

Mutti, F.G., and Kroutil, W. 2012. Asymmetric bio-amination of ketones in organic solvents. *Adv. Synth. Catal.* 354: 3409–3413.

Pace, V., Hoyos, P., Castoldi, L., Domínguez de María, P., and Alcántara, A.R. 2012. 2-Methyltetrahydrofuran (2-MeTHF): a biomass-derived solvent with broad application in organic chemistry. *ChemSusChem* 5: 1369–1379.

Pang, K., Hou, Y., Wu, W., Guo, W., Peng, W., and Marsh, K.N. 2012. Efficient separation of phenols from oils via forming deep eutectic solvents. *Green Chem.* 14: 2398–2401.

Pérez-Sánchez, M., Sandoval, M., Hernáiz, M.J., and Domínguez de María, P. 2013. Biocatalysis in biomass-derived solvents: the quest for fully sustainable chemical processes. *Curr. Org. Chem.* 17: 1188–1199.

Rastogi, R.P., and Bassi, P.S. 1964. Mechanism of eutectic crystallization. *J. Phys. Chem.* 68: 2398–2406.

Reetz, M.T. 2011. Laboratory evolution of stereoselective enzymes: a prolific source of catalysts for asymmetric reactions. *Angew. Chem. Int. Ed.* 50: 138–174.

Reetz, M.T. 2012. Laboratory evolution of stereoselective enzymes as a means to expand the toolbox of organic chemistrs. *Tetrahedron* 68: 7530–7548.

Rose, M., and Palkovits, R. 2012. Isosorbide as a renewable platform chemical for versatile applications—quo vadis? *ChemSusChem* 9: 167–176.

Ruß, C., and König, B. 2012. Low melting mixtures in organic synthesis—an alternative to ionic liquids? *Green Chem.* 14: 2969–2982.

Savile, C.K., Janey, J.M., Mundorff, E.C., Moore, J.C., Tam, S., Jarvis, W.R., Colbeck, J.C., et al. 2010. Biocatalytic asymmetric synthesis of chiral amines from ketones applied to sitagliptin manufacture. *Science* 329: 305–309.

Shahbaz, K., Mjalli, F.S., Hashim, M.A., and AlNashef, I.M. 2010. Using deep eutectic solvents for the removal of glycerol from palm oil-based biodiesel. *J. App. Sci.* 10: 3349–3354.

Shahbaz, K., Mjalli, F.S., Hashim, M.A., and AlNashef, I.M. 2011a. Using deep eutectic solvents based on methyl triphenyl phosphonium bromide for the removal of glycerol from palm-oil-based biodiesel. *Energy Fuels* 25: 2671–2678.

Shahbaz, K., Mjalli, F.S., Hashim, M.A., and AlNashef, I.M. 2011b. Eutectic solvents for the removal of residual palm oil-based biodiesel catalyst. *Sep. Purif. Technol.* 81: 216–222.

Truppo, M.D., Strotman, H., and Hughes, G. 2012. Development of an immobilized transaminase capable of operating in organic solvent. *ChemCatChem* 4: 1071–1074.

van Putten, R.J., van der Waal, J.C., de Jong, E., Rasrendra, C.B., Heeres, H.J., and de Vries, J.G. 2013. Hydroxymethylfurfual, a versatile platform chemical made from renewable resources. *Chem. Rev.* DOI: 10.1021/cr300182k.

van Spronsen, J., Witkamp, G.J., Hollmann, F., Choi, Y.H., and Verpoorte, R. 2011. Process for extracting materials from biological material. WO2011/155829.

Wasserscheid, P., and Welton, T. 2007. *Ionic liquids in synthesis*. Wiley-VCH, Weinheim.

Welton, T. 1999. Room-temperature ionic liquids. *Chem. Rev.* 99: 2071–2084.

Whittall, J., and Sutton, P. 2009, 2012. *Practical methods for biocatalysis and biotransformations*. 2 vols. Wiley-VCH, Weinheim.

Zhang, Q., De Oliveria Vigier, K., Royer, S., and Jerome, F. 2012. Deep eutectic solvents: syntheses, properties and applications. *Chem. Soc. Rev.* 41: 7108–7146.

Zhao, H., and Baker, G.A. 2013. Ionic liquids and deep eutectic solvents for biodiesel synthesis: a review. *J. Chem. Technol. Biotechnol.* 88: 3–12.

Zhao, H., Baker, G.A., and Holmes, S. 2011a. New eutectic ionic liquids for lipase activation and enzymatic preparation of biodiesel. *Org. Biomol. Chem.* 9: 1908–1916.

Zhao, H., Baker, G.A., and Holmes, S. 2011b. Protease activation in glycerol-based deep eutectic solvents. *J. Mol. Cat. B Enzym.* 72: 163–167.

Zhao, H., Zhang, C., and Crittle, T.D. 2013. Choline-based deep eutectic solvents for enzymatic preparation of biodiesel from soybean oil. *J. Mol. Cat. B Enzym.* 85–86: 243–247.

chapter five

Synthesis and applications of ionic liquids as pharmaceutical materials

Hua Zhao and Sanjay V. Malhotra

Contents

5.1 Introduction

Ionic liquids (ILs) consist of ions and remain liquid at temperatures below 100°C. The so-called room temperature ionic liquids (RTILs) are most suitable as solvents for chemical reactions and other applications. Comparing with conventional organic solvents, ILs have many favorable properties, such as very low vapor pressure, a wide liquid range, low flammability, high ionic conductivity, high thermal conductivity, high dissolution power toward many substrates, high thermal and chemical stability, and a wide electrochemical potential window.[1] Due to these unique properties, ILs have been widely recognized as solvents or reagents in a variety of applications, including organic catalysis,[1–9] inorganic synthesis,[10] biocatalysis,[8,11–16] polymerization,[17,18] and engineering fluids.[19–21] Typical IL cations are nitrogen containing (such as alkylammonium, *N,N'*-dialkylimidazolium, *N*-alkylpyridinium, and pyrrolidinium) or phosphorous containing (such as alkylphosphonium). The common choices of anions include halides, BF_4^-, PF_6^-, $CH_3CO_2^-$, $CF_3CO_2^-$, NO_3^-,

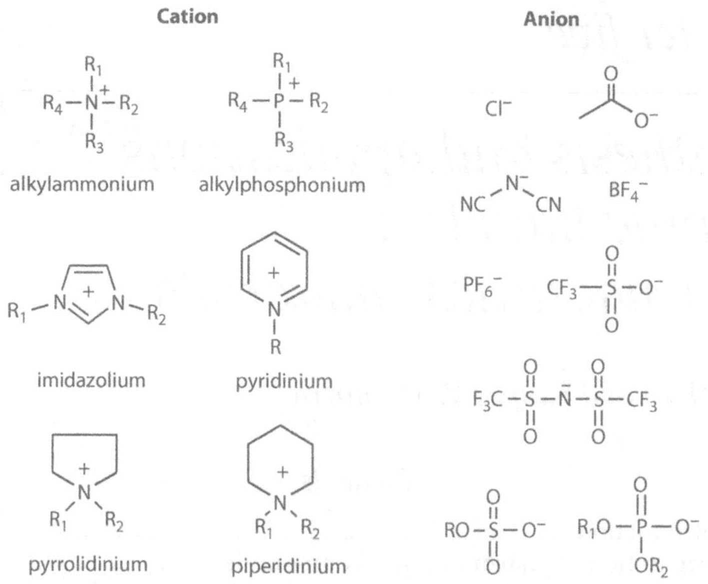

Scheme 5.1a Structures of common IL cations and anions.

Tf_2N^- (i.e., $(CF_3SO_2)_2N^-$), $[RSO_4]^-$, $[R_2PO_4]^-$, etc. Some typical cations and anions are illustrated in Scheme 5.1a.

Due to their designable properties, ILs have recently been exploited as solvents or materials for a variety of pharmaceutical applications. Although there are several articles[22–24] reviewing different aspects of this subject, this chapter provides the most up-to-date discussion of the solvent role of ILs in drug molecule preparations, the carrier role of ILs in drug delivery, the active pharmaceutical ingredients (APIs) role of ILs, and other applications.

5.2 *Preparation of pharmaceutical compounds in ILs*

ILs have many favorable properties to replace volatile organic solvents for the synthesis of pharmaceutical compounds. In the examples illustrated below, the use of ILs often led to higher yields, better selectivity, or simple product isolation. The reaction systems involving ILs could be homogeneous or heterogeneous phases, enabling a higher flexibility in manipulating the reactions. In addition, ILs are suitable for both chemical and enzymatic preparations of drug molecules. Due to the large volume of literature on the synthesis of drug-related molecules in ILs, this chapter only discusses some of the most relevant and typical examples of these reactions.

5.2.1 Synthesis of anti-inflammatory and analgesic drugs

2-Arylpropionic acids belong to an important class of anti-inflammatory and analgesic drugs. The Dupont group[25] discovered that [RuCl$_2$-(S)-BINAP]$_2$·NEt$_3$ catalyst precursor could be dissolved in [BMIM][BF$_4$] to hydrogenate 2-arylacrylic acids (aryl = Ph or 6-MeO-naphthyl) (Scheme 5.1b); the enantioselectivities obtained were similar to or higher than those from homogeneous media. In addition, hydrogenated 2-arylpropionic acids in isopropanol could be separated by simple decantation and the ionic catalyst solution could be reused without much loss of activity and selectivity.

As nonsteroidal anti-inflammatory drugs (NSAIDs), pravadoline is found to be an agonist of the cannabinoid receptor, and thus pravadoline and its analogues have been used as a probe for neurochemical receptors.[26] Earle et al.[27] developed a two-step synthetic route in [BMIM][PF$_6$]: a base-catalyzed alkylation and a Friedel–Crafts acylation without the use of Lewis acids (Scheme 5.2). The overall isolated yield was 90–94%.

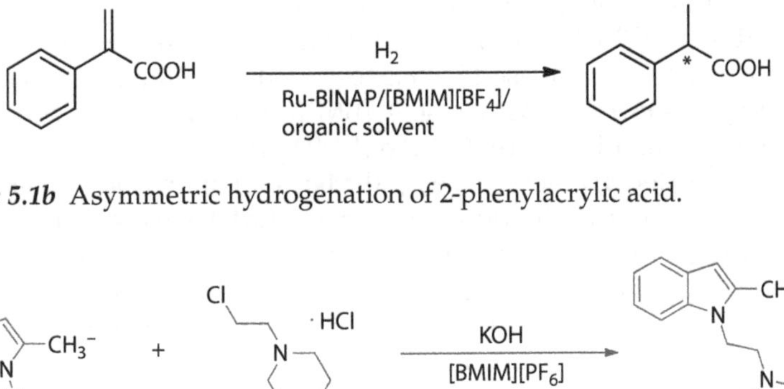

Scheme 5.1b Asymmetric hydrogenation of 2-phenylacrylic acid.

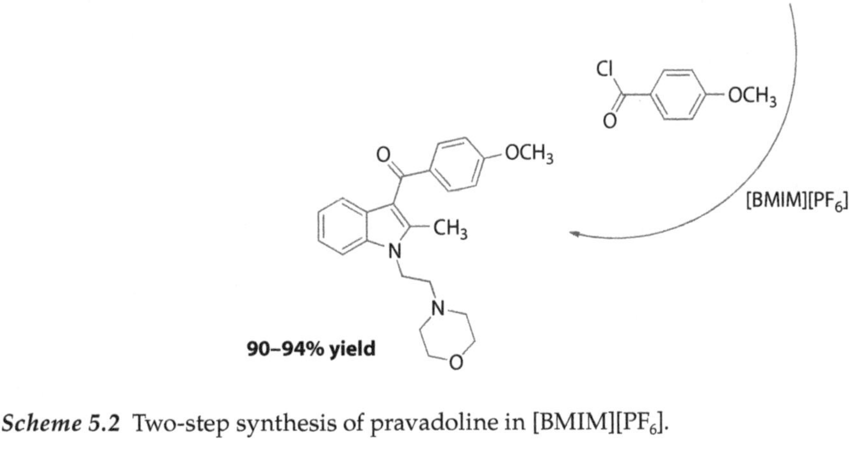

Scheme 5.2 Two-step synthesis of pravadoline in [BMIM][PF$_6$].

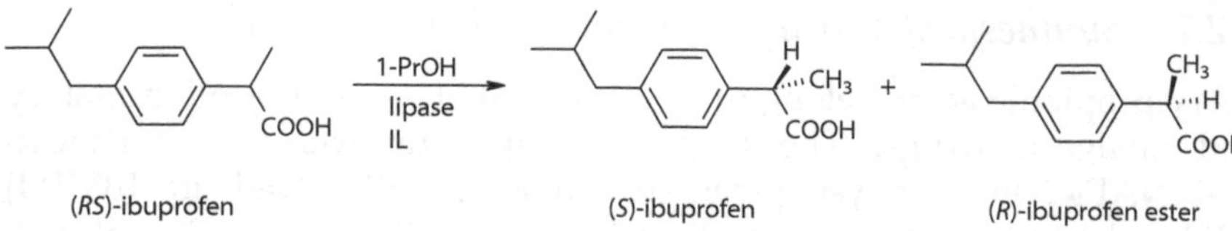

Scheme 5.3 Enantioselective esterification of (*RS*)-ibuprofen with 1-propanol.

S-(+)-Ibuprofen has been shown to be 100 times more biologically active than the *R*-(–) form as the anti-inflammatory drug. There are different approaches involving ILs to resolve the ibuprofen racemic mixture. Through an enzymatic resolution method, Yu et al.[28] carried out *Candida rugosa* lipase-catalyzed esterification of ibuprofen with 1-propanol in several ILs (Scheme 5.3) and found the enantioselectivity in [BMIM][PF_6] (E = 24.1) was about twice that in isooctane (E = 13.0). In addition, the lipase stability was improved by 25% in [BMIM][PF_6] compared with that in isooctane.

In another report,[29] several lipases were examined for the same resolution reaction (Scheme 5.3) in a biphasic system of isooctane and [BMIM][PF_6] (or [BMIM][BF_4]). The best enzyme in terms of enantioselectivity was *C. rugosa* lipase (E = 8.5, ee_s = 60%), followed by the lipase from *Aspergillus niger* AC-54 (E = 4.6). [BMIM][PF_6] was more effective in improving the enantioselectivity than [BMIM][BF_4].

The supported IL membranes (SILMs) offer many advantages over conventional supported liquid membranes (SLMs), such as (1) minimum loss of impregnated liquid in the membrane due to vaporization, (2) high carrier loadings, (3) tunable properties of the membrane solvent, (4) high stability and flux rates, and (5) improved selectivity of the liquid membrane.[30] Miyako et al.[31] demonstrated the lipase-facilitated transport of (*S*)-ibuprofen through an SILM. Lipase from *C. rugosa* (CRL) was used in the feed phase (interface 1) to selectively convert (*S*)-ibuprofen to corresponding (*S*)-ester, and (*R*)-ibuprofen was collected from the feed phase; the (*S*)-ester was then dissolved in the IL phase of SILM and diffused through the membrane. In the receiving phase (interface 2), lipase from porcine pancreas (PPL) hydrolyzed the (*S*)-ester to produce the water-soluble (*S*)-ibuprofen. That study examined the effect of different ILs and organic solvents as the liquid membranes in SILMs. The SILMs based on ILs (PF_6^- and Tf_2N^-) enabled a higher initial permeate flux of ibuprofen than those based on organic solvents (C_5-C_{12} alkanes), with [BMIM][Tf_2N] being selected as the best IL. At 48 h of operation, (*R*)-ibuprofen with 75.1% ee was obtained in the feed phase using CRL, and (*S*)-ibuprofen with 75.1% ee was collected in the receiving phase using PPL.

Another interesting development is the coupling of SILM with microfluidic devices for the selective and rapid separation of racemic mixtures at the analytical scale.[32] Huh et al.[32] illustrated the configuration of the

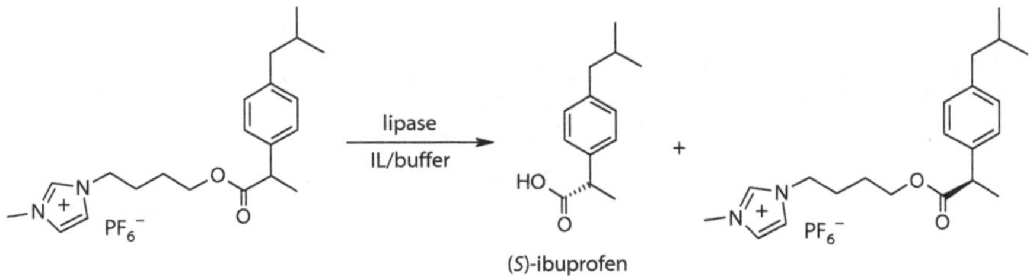

Scheme 5.4 Lipase-catalyzed kinetic resolution of IL-anchored ibuprofen ester.

microchannel with a three-phase flow, and the transport of (*R*, *S*)-ibuprofen through the IL flow (ILF) in the microchannel. The involved reactions and selective diffusion of (*R*, *S*)-ibuprofen are similar to the earlier discussion, where (*S*)-ibuprofen was selectively esterified and transported through the membrane via ILF. At an IL flow rate of 0.30 ml/h, (*S*)-ibuprofen with 78% ee was detected in the receiving phase, and (*R*)-ibuprofen with 48% ee was obtained in the feed phase. Although the SILM system could transport more solutes from the feed phase to the receiving phase given sufficient reaction time (20–40 h), the microfludic system affords a fast and selective separation of solutes (~30–60 s) at the analytical scale.

Interestingly, some studies incorporated the IL structure into the substrates for improved enzymatic resolutions. For example, Naik et al.[33] anchored ibuprofen with a hydroxyl group appended IL, and hydrolyzed the anchored ibuprofen by lipases in a 50/50 (v/v) mixture of organic solvent (or IL) and 0.1 M phosphate buffer (Scheme 5.4). When the reaction in dimethyl sulfoxide (DMSO)/buffer was catalyzed by *Candida antarctica* lipase B (CALB), (*S*)-ibuprofen was obtained with 86% ee and the isolated yield of 87% of theory. The same reaction in [BMIM][PF_6]/buffer catalyzed by CALB produced 80% ee and the isolated yield of 80% of theory.

5.2.2 Synthesis of antitumor and antiviral drugs

Boronated α-amino acids such as L-*p*-dihydroxyborylphenylalanine (L-BPA) can be used for the boron neutron capture therapy (BNCT). The cross-coupling of protected *p*-iodophenylalanine with pinacolborane was carried in different ILs (Scheme 5.5), resulting in 82–89% yields after 20 min.[34] The product was easily isolated by the separation of phases, and the catalyst remained in the IL layer. Kurata et al.[35] performed the transesterification of methyl caffeate to synthesize caffeic acid phenethyl ester analogues catalyzed by *C. antarctica* lipase B (CALB) in [BMIM][Tf_2N]. In particular, 2-cyclohexylethyl caffeate and 3-cyclohexylpropyl caffeate have shown strong antiproliferative activities that are comparable to 5-fluorouracil using MTT (3-(4,5 dimethyl-2-thazolyl)-2,5-diphenyl-2H tetrazolium bromide) assay.

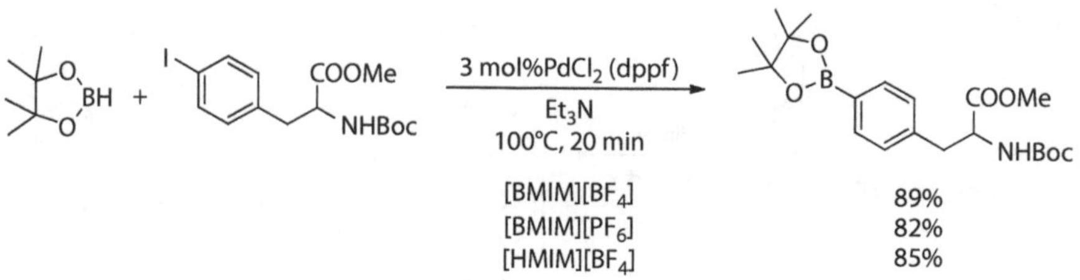

Scheme 5.5 Cross-coupling of protected *p*-iodophenylalanine with pinacolborane.

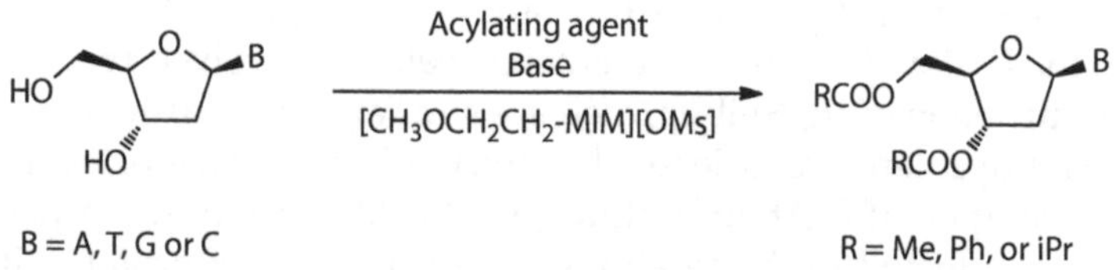

Scheme 5.6 Acylation of 2′-deoxyribonucleosides in IL.

Nucleoside analogues are known for applications in cancer and viral chemotherapy. Uzagare et al.[36] noticed that thymidine and other 2′-deoxynucleosides were better dissolved by 1-methoxyethyl-3-methylimidazolium methanesulfonate [$CH_3OCH_2CH_2$-MIM][OMs] than conventional solvents and other methanesulfonate ILs ([EMIM]$^+$ and [BMIM]$^+$); therefore, they further investigated the acylation and peracylation of 2′-deoxynucleosides using different acylating agents and bases (Scheme 5.6), achieving 85–95% yields. Liu et al.[37] prepared polymerizable esters of nucleoside drugs through CALB-catalyzed regioselective transesterification in a mixture of 90% acetone and 10% [BMIM][BF_4], resulting in eight times higher yields and three times faster reaction rates (Scheme 5.7). 5-Halouracil nucleosides have great pharmaceutical interests due to their antineoplastic and antiviral properties. The preparations of 5-halo derivatives of both protected and unprotected uridine and 2′-deoxyuridine were conducted in various IL media (i.e., [MeOEt-MIM][$MeSO_3$], [MeOEt-MIM][CF_3COO], [BMIM][$MeSO_3$], and [BMIM][CF_3COO]), leading to moderate to high yields in many cases (Scheme 5.8).[38]

5.2.3 Synthesis of other drug-related molecules

Imidazo[1,2-α]pyridines represent an important category of pharmaceutical compounds with various biological activities. Shaabani et al.[39] prepared 3-aminoimidazo[1,2-α]pyridines through multicomponent condensation of an aldehyde, 2-amino-5-methylpyridine (or 2-amino-5-bromopyridine),

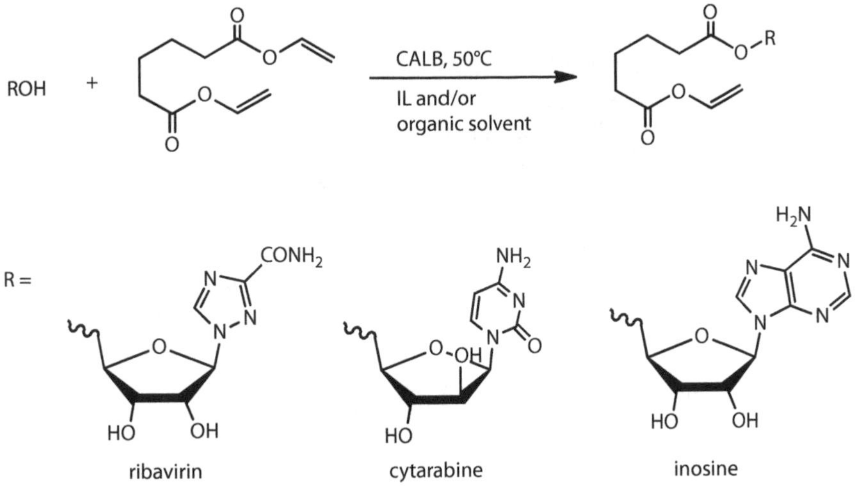

Scheme 5.7 Enzymatic regioselective synthesis of vinyl ribavirin esters, vinyl cytarabine esters, and vinyl inosine esters.

and an isocyanide using [BMIM]Br (Scheme 5.9) and achieved good to excellent yields (70–99%) at room temperature.

Thiazolidinone and its derivatives have a wide range of biological activities, including anti-inflammatory, antiproliferative, anticyclooxygenase, antihistaminic, and antibacterial activities. In particular, as nonnucleoside HIV-I reverse transcriptase inhibitors (NNRTIs), some 2,3-diaryl-1,3-thiazolidin-4-ones are very effective against the HIV-1 replication. Zhang et al.[40] reported an effective synthesis of 2,3-disubstituted-1,3-thiazolidin-4-one derivatives by a three-component reaction of aldehyde, amine, and mercaptoacetic acid in [BMIM][PF_6] (Scheme 5.10). Some derivatives exhibited moderate activities against trypomastigote forms of *Trypanosoma brucei.*

As a key intermediate in the synthesis of chiral drug (*S*)-lubeluzole, (*R,S*)-1-chloro-3-(3,4-difluorophenoxy)-2-propanol (*rac*-CDPP) was enzymatically resolved by the transesterification of this racemic mixture with vinyl butyrate catalyzed by lipase from *Pseudomonas aeruginosa* in hexane with [BMIM][PF_6] or [BMIM][BF_4] as the cosolvent (Scheme 5.11).[41] A high conversion (>49%) and enantiomeric excess (ee > 99.9%) was obtained in 6 h at 30°C when using [BMIM][PF_6] as cosolvent in a two-phase system. Generally, a higher enzyme activity was seen in [BMIM][PF_6] than in [BMIM][BF_4].

Silver nanoparticles and silver/polystyrene core-shell nanoparticles were prepared with a high uniformity in [BMIM][BF_4].[42] The use of this IL minimized the nanoparticle aggregation during the preparation step.

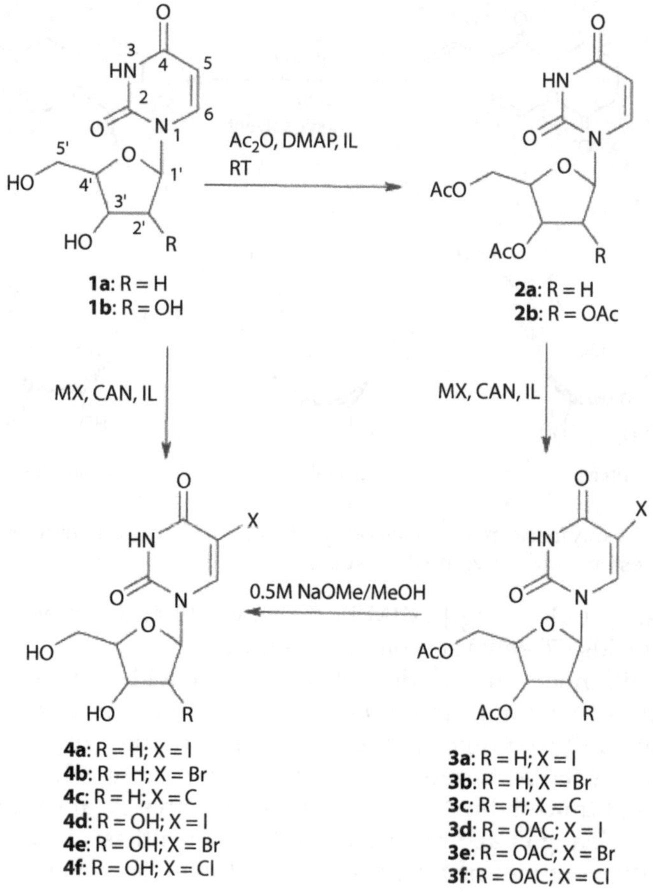

Scheme 5.8 Synthesis of 5-halouridines and 5-halo-2′-deoxyuridines.

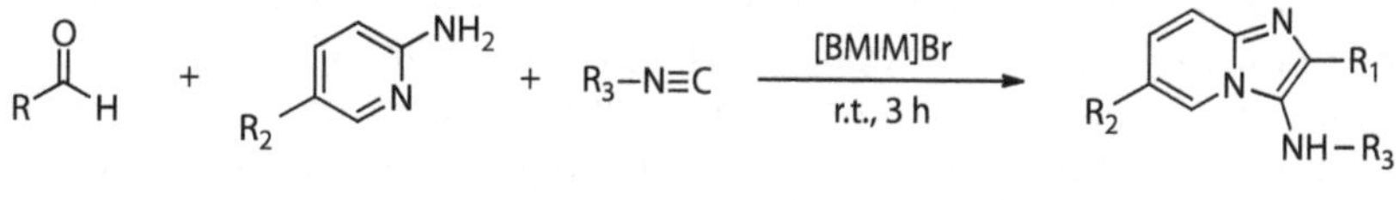

Scheme 5.9 Condensation synthesis of 3-aminoimidazo[1,2-α]pyridines.

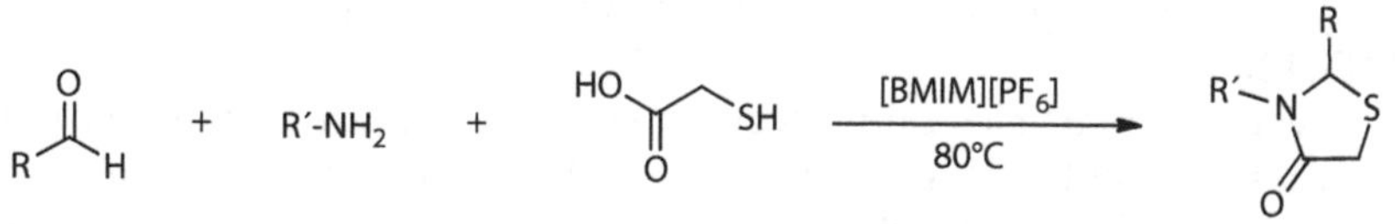

Scheme 5.10 Synthesis of 2,3-disubstituted-1,3-thiazolidin-4-one derivatives.

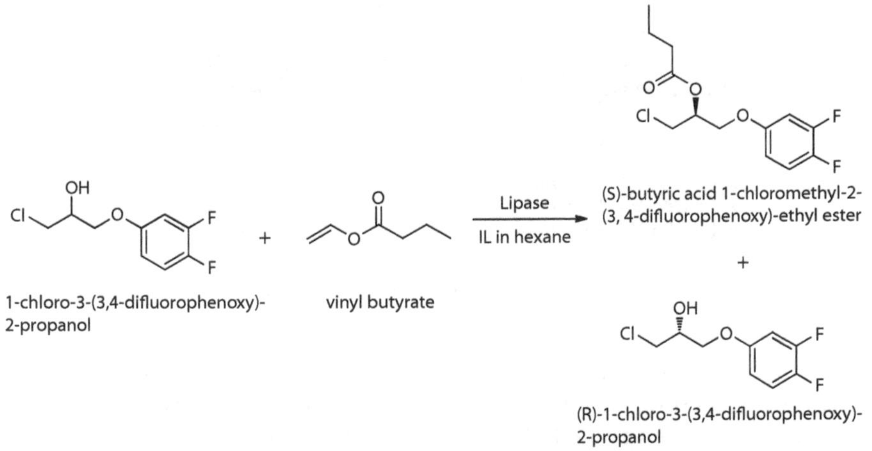

Scheme 5.11 Enzymatic transesterification of *rac*-CDPP to (*S*)-butyric acid 1-chloromethyl-2-(3,4-difluoro-phenoxy)-ethyl ester using IL as cosolvent.

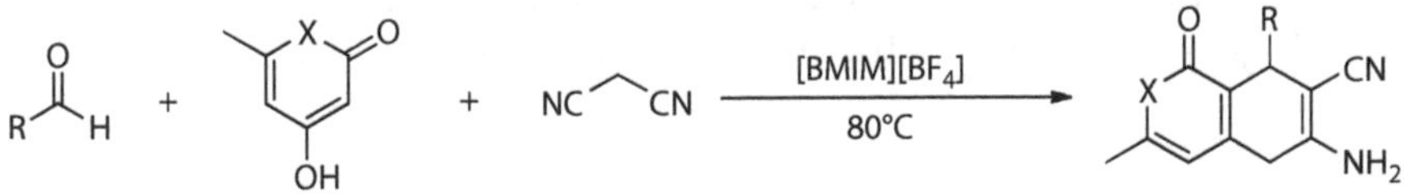

Scheme 5.12 Synthesis of pyrano[3,2-*c*]pyridone and pyrano[4,3-*b*]pyran.

Both silver nanoparticles and core-shell nanoparticles exhibit high antimicrobial activities against *Escherichia coli* and *Staphylococcus aureus.*

Fan et al.[43] prepared pyrano[3,2-*c*]pyridone and pyrano[4,3-*b*]pyran derivatives through a multicomponent reaction of aldehyde, 4-hydroxypyridin-2(1*H*)-one or 4-hydroxy-2-pyranone, and malononitrile in [BMIM][BF_4] (Scheme 5.12). Following a similar strategy, they prepared a series of pyrimidine nucleoside-pyrano[3,2-*c*]pyridone or pyrano[4,3-*b*]pyran hybrids, showing high antiviral and antileishmanial activities.

(*S*)-3-Chloro-1-phenyl-1-propanol is a known chiral synthon for the synthesis of antidepressant drugs such as fluoxetine, tomoxetine, and nisoxetine. Choi et al.[44] suggested that the [BMIM][Tf_2N]/water two-phase system was able to improve the substrate solubility and enzyme inhibition during the enantioselective reduction of 3-chloro-1-phenyl-1-propanone into (*S*)-3-chloro-1-phenyl-1-propanol (ee > 99%) catalyzed by yeast reductase YOL151W.

Talisman et al.[45] found that *O*-glycosidation of glycosyl bromides with therapeutically relevant acceptors (Scheme 5.13) could be effectively facilitated by silver *N*-heterocyclic carbene (Ag-NHC) complexes derived from ILs: 1-benzyl-3-methylimidazol-2-ylidene silver (I) chloride (**1**) and 1-(2-methoxyethyl)-3-methylimidazol-2-lylidene silver (I) chloride (**2**).

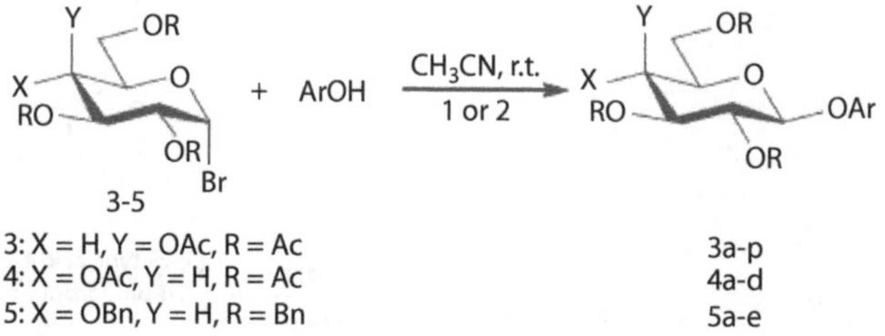

Scheme 5.13 *O*-Glycosidation of different sugar donors and phenolic acceptors using Ag-NHC complexes.

This method was further explored to synthesize a series of novel glycosides using steroid, flavone, and coumarin scaffolds; all reactions were highly selective and yielded the β-products regardless of neighboring group effects.

Dandia and Jain[46] carried out a new multicomponent reaction (MCR) of urea (or thiourea), aryl aldehydes, and 3-methyl-1-phenyl-2-pyrazolin-5-one in $[BMIM][BF_4]$, producing a whole set of spiropyrimidine heterocycles with high yields, but not the expected Biginelli product (Scheme 5.14).

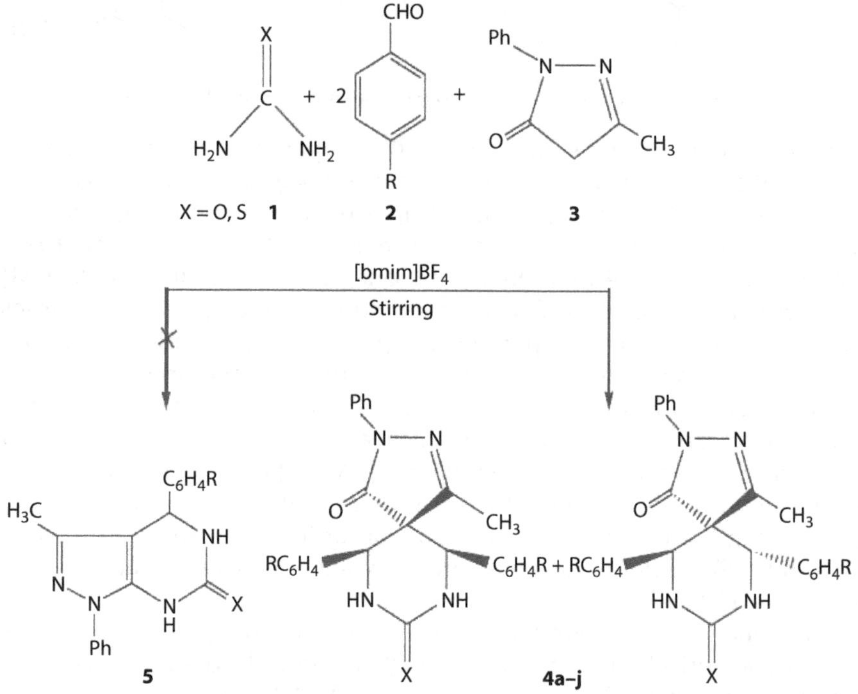

Scheme 5.14 Synthesis of 7,7-dimethyl-4-phenyl-4,6,7,8-tetrahydroquinazoline-2,5-dione.

They further examined these compounds by in vitro antimicrobial screening against a panel of pathogenic strains of bacteria and fungi, and found some of the compounds to be equipotent or more potent than the commercial antibiotics.

5.3 IL-assisted drug delivery

Many sparingly soluble drugs (such as acyclovir, albendazole, danazol, acetaminophen, and caffeine) can have a higher solubility in common ILs.[47,48] This has prompted the development of new drug delivery systems using ILs.

Cellulose–heparin composite fibers were prepared by dissolving cellulose and heparin in [BMIM]Cl and [BMIM][benzoate], respectively, prior to the mixing of these two solutions, and by further electrospinning to form micro- and nano-sized fibers (see Figure 5.1).[49] After the fiber formation, ILs were removed from the biopolymer fiber using ethanol. Anticoagulant activity was observed for the cellulose-heparin composite fibers, suggesting the bioactivity of heparin maintained after the electrospinning process. Using a similar IL dissolution approach, but without electrospinning, the same group[50] constructed heparin–cellulose–charcoal

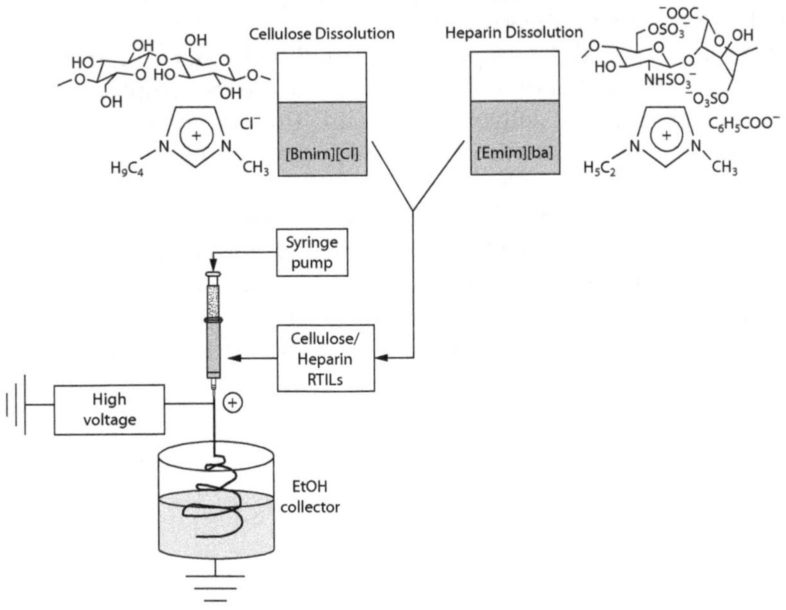

Figure 5.1 Preparation of cellulose–heparin composite fibers using ILs by electrospinning. (Reprinted with permission from Viswanathan, G. et al., *Biomacromolecules*, 7, 2006, 415–418. Copyright © 2006 by the American Chemical Society.)

composites to form activated charcoal beads with high biocompatibility and blood compatibility, as well as the decreased size of active pores. These new biomaterials have potential applications for direct hemoperfusion to remove free-diluted and protein-bound toxins of small size, or as oral agents if strict preservation of large molecules such as proteins is needed.

Tsioptsias and Panayiotou[51] prepared cellulose-nanohydroxyapatite composite scaffolds with high and open porosity by a poly(methyl methacrylate) (PMMA) particulate leaching technique. They observed a better dispersion of hydroxyapatite in cellulose using [BMIM]Cl than that using DMA + LiCl as the solvent. The cellulose-hydroxyapatite matrix was used to encapsulate an antibiotic drug (i.e., amoxicillin).

Viau et al.[52] has demonstrated that ionogels are effective drug release systems for imidazolium ibuprofenate (Scheme 5.15), and the drug release was controlled by the inner and the outer surface of the ionogel, and the chemical nature of the surface. The IL form avoids the issue of drug polymorphism.

Pang et al.[53] synthesized porous and hollow silica particles from sodium silicate through a straightforward acid gelation method in [BMIM][BF_4]. They further studied the drug release behavior of ibuprofen from porous silica particles into the simulated intestinal fluid and the simulated gastric fluid, respectively, implying the potential use of these silica particles in the drug delivery system.

Mahkam et al.[54] prepared IL-modified silica nanoparticles (Scheme 5.16) through a two-step strategy, and further employed them in the pH-responsive oral delivery of insulin. Trapped insulin in nanoparticles was studied for its release into two enzyme-free simulated gastric (SGF, pH 1) and intestinal (SIF, pH 7.4) fluids. It was found that in physiological buffer solution (pH 7.4), the deprotonation of silanol groups causes a partial negative surface charge on the modified silica nanoparticle, which induces a slow release of insulin due to the strong electrostatic repulsion.

An antisolvent precipitation technique using an IL was explored aiming to improve the bioavailability of poorly soluble drugs[55]: rifampicin was

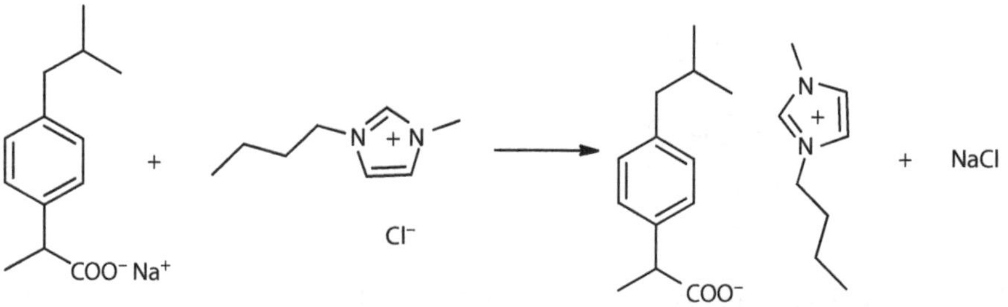

Scheme 5.15 Preparation of imidazolium ibuprofenate.

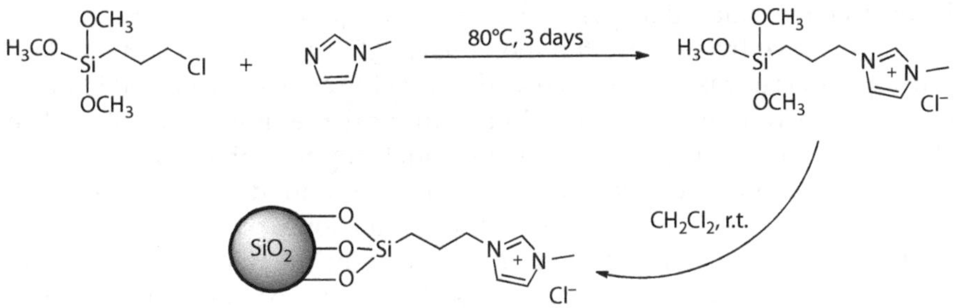

Scheme 5.16 Preparation of IL-functionalized silica nanoparticles.

first dissolved in 1-ethyl-3-methylimidazolium methylphosphonate, and the IL solution was introduced into a phosphate buffer (as an antisolvent) to form particles in the submicron range (280–360 nm) with or without hydroxypropyl methylcellulose as the stabilizer; the amorphous ultrafine particles exhibited a higher solubility and a faster dissolution rate.

Tsai et al.[56] employed [BMIM][BF_4] as an electrolyte to electrochemically fabricate nano Pd-Au particles onto electrode surfaces to construct a glassy carbon electrode. The new electrode exhibited a high electrochemical activity and stability in various pH solutions, and could be used for the detection of epinephrine, dopamine, and uric acid in pH 7.0 phosphate buffer solution.

ILs can increase the solubility of sparingly soluble drugs and improve their topical and transdermal delivery. The Goto group[57] developed IL-in-oil microemulsions that contain drug molecules (such as acyclovir) in the IL core and the continuous oil phase for easy topical or transdermal transport (Figure 5.2). The formation of nanometer-sized IL droplets in

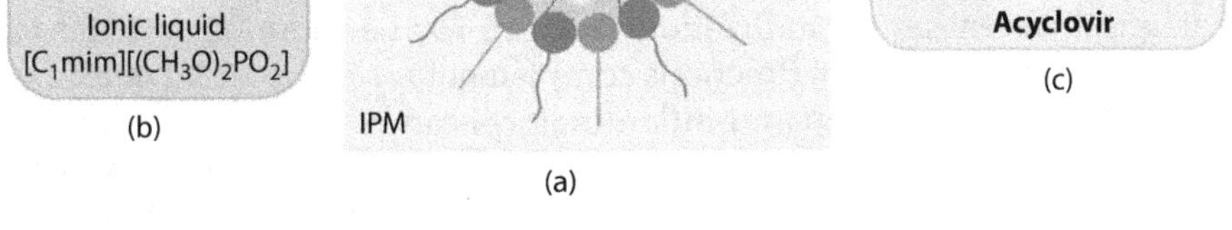

Figure 5.2 Illustrations of (a) IL-in-oil (IL/o) microemulsions containing drug molecules, (b) IL structure, and (c) structure of acyclovir. (Reprinted with permission from Moniruzzaman, M. et al., *Int. J. Pharm.*, 400, 2010, 243–250. Copyright © 2010 by Elsevier.)

isopropyl myristate (IPM) was achieved through mixing nonionic surfactants, polyoxyethylene sorbitan monooleate (Tween-80), and sorbitan laurate (Span-20); it was also found that ILs carrying coordinating and H-bond basic anions were most effective in forming microemulsion droplets. The IL-in-oil microemulsions have shown much higher solubilities for drug molecules, including acyclovir, methotrexate, and dantrolene sodium, than IL free micelles, water-in-oil microemulsions, or water. The same group[58] further evaluated the drug release from the new microemulsion system in vitro through full-thickness skin pieces of Yucatan micropigs (YMPs) and observed that the use of IL-in-oil microemulsions improved the skin permeability of acyclovir by several orders of magnitude and also induced the transdermal permeation of drug molecules. They further examined the cytotoxicity of a new carrier and found microemulsions containing 4 wt% IL retained 80% cell viability compared to Dulbecco's phosphate-buffered salines, implying a low cytotoxicity of the new delivery system. However, they also point out that nontoxic ILs need to be developed for the drug delivery. In a separate study, Dobler et al.[59] modified the conventional oil-in-water (O/W) and water-in-oil (W/O) emulsions by replacing the water phase with hydrophilic [HMIM]Cl and replacing the oil phase with hydrophobic [BMIM][PF_6]. The resulting new emulsions showed antimicrobial activity (with application as preservatives), and low cytotoxicity of the carriers (with application for drug delivery). In addition, for lipophilic substances in [BMIM][PF_6]-containing emulsion, a more efficient penetration into the deeper skin layers was observed.

5.4 ILs as active pharmaceutical ingredients (APIs)

For many decades the pharmaceutical industry has successfully employed the approach of combining salts of two drugs to achieve the combined pharmacological effects.[60,61] Recently, similar effort has been touted as ionic liquid-active pharmaceutical ingredients (IL-API).[62,63] In its simplest manifestation (and as common practice in the pharmaceutical industry), the idea is to transform an API into an ionic form followed by exchange of the counter ion (metathesis) of another API. Thereby, rendering the salt (ion pair) as a liquid. The counter ion not only serves to frustrate ionic packing (preventing crystallization) but can furthermore be selected to display desired biological functions complementary to the API. For example, sodium ibuprofen (an anti-inflammatory) can be paired with didecyldimethylammonium bromide (antibacterial and anti-inflammatory) to yield the IL-API didecyldimethylammonium ibuprofenate, which retains dual biological roles.[63] The Scott group[64] further applied the solid-state structures of the crystalline salts as a basis to study an anticrystal engineering approach to the preparation of pharmaceutically active ILs.

Betulinic acid (3β-hydroxy-lup-20(29)-en-28-oic acid, **1**) is a natural pentacyclic lupane-type triterpene (Scheme 5.17) that exists in various plants, including birch trees. This compound and its derivatives possess many favorable biological properties such as anticancer, anti-HIV-1 (human immunodeficiency virus type 1), antibacterial, antimalarial, anti-inflammatory, and anthelmintic activities.[65–70] However, this natural compound is poorly soluble in water (only about 0.02 μg ml^{-1} at room temperature).[71] To overcome this bottleneck, a number of ionic derivatives of betulinic acid (Scheme 5.17) were prepared, and it was found that these derivatives are better inhibitors for HIV-1 protease[72] and several cancer cell lines (such as melanoma A375, neuroblastoma SH-SY5Y, and breast adenocarcinoma MCF7).[73]

Ferraz et al.[74] adopted the anion exchange resin method to prepare several IL forms of β-lactam antibiotics ampicillin where ampicillin serves as counter anion (Scheme 5.18). Most of these salts have melting points under 100°C and decomposition temperatures greater than 210°C. In particular, [cholinium][ampicillin] is a promising IL-API with low melting point (58.0°C), very high water solubility (miscible), high biocompatibility, and low toxicity of the cholinium cation.

5.5 Miscellaneous applications

Liu and Jiang[75] studied [BMIM][BF_4] as a matrix medium for the determination of residual solvents (such as acetonitrile, dichloromethane, *N*-methyl-2-pyrrolidone (NMP), toluene, dimethylformamide (DMF), and *n*-butyl ether) in pharmaceuticals (i.e., adefovir dipivoxil) by static headspace gas chromatography and found this IL increased the sensitivity of higher boiling point analytes and was a more environmentally benign alternative to organic matrix media such as DMSO and DMF. 1-Hexylpyridinium hexafluorophosphate was employed as a microextraction solvent to concentrate terazosin (an effective drug for hypertension and benign prostatic hyperplasia) prior to spectrofluorimetric quantification of terazosin concentration.[76] This method has a low limit of detection (LOD) of 0.027 μg L^{-1} and a relative standard deviation (RSD) of 2.4%. Bratkowska et al.[77] examined two imidazolium supported IL phases (SILPs) containing different anions (i.e., $CF_3SO_3^-$ and BF_4^-) as solid-phase extraction sorbents for removing acidic pharmaceuticals (such as salicylic acid, naproxen, fenoprofen, diclofenac, ibuprofen, and gemfibrozil) from aqueous samples under strong anion exchange conditions. In particular, they found the SILP materials based on [MI^+][CF_3COO^-] have a high selectivity and capacity, which is comparable to the commercially available Oasis MAX sorbent.

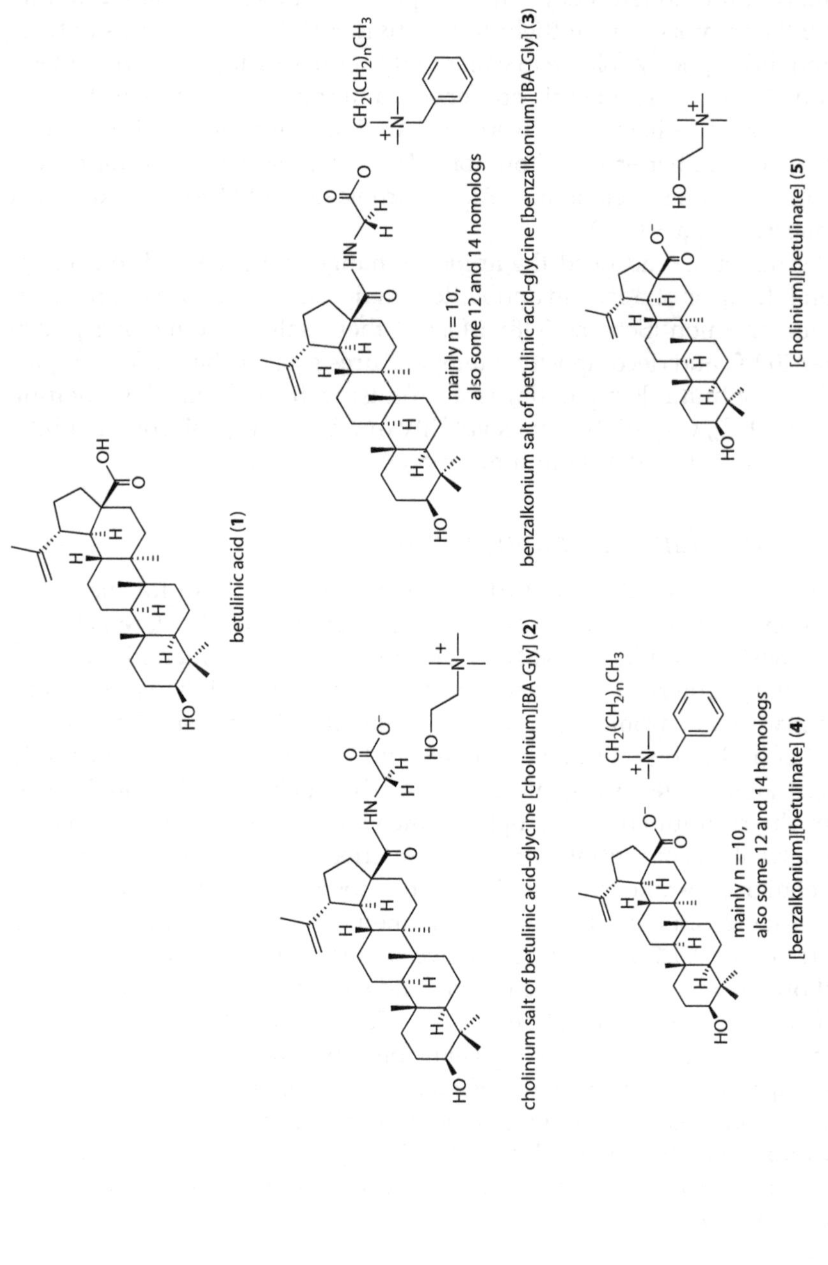

Scheme 5.17 Betulinic acid and its ionic derivatives.

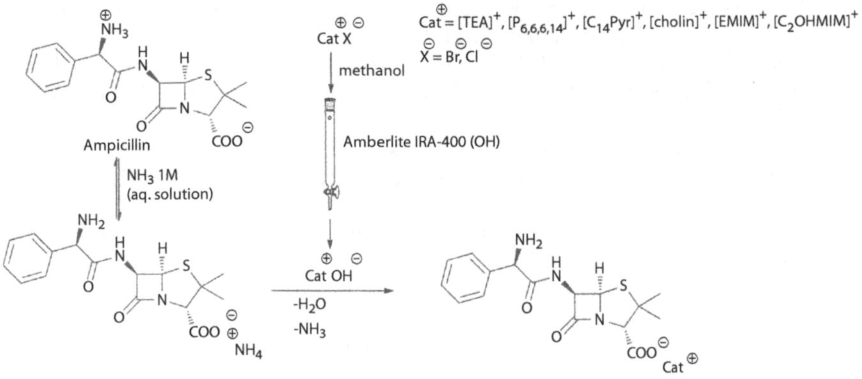

Scheme 5.18 Synthesis of ampicillin-based ILs.

5.6 *Summary*

As demonstrated by examples in this chapter, ILs can be suitable solvents or materials for a variety of pharmaceutical applications. The tunable solvent properties of ILs enable new chemical/enzymatic synthetic routes or higher yields and selectivities in drug synthesis; the high dissolution power of ILs allow new methods for drug delivery; more excitingly, active pharmaceutical ingredients themselves can be made into ILs to avoid drug polymorphism problems. The main challenges in the field include the high cost of most ILs, as well as the development of less toxic and more biodegradable types of ILs.

References

1. Wasserscheid, P., Welton, T. *Ionic Liquids in Synthesis.* 2nd ed. Wiley-VCH, Weinheim, 2008.
2. Gordon, C. M. New developments in catalysis using ionic liquids. *Appl. Catal. A Gen.* 2001, 222, 101–117.
3. Houlton, S. Ionic liquids: the route to cleaner and more efficient fine chemical synthesis? *Chem. Week.* 2004, February 25, s10–s11.
4. Seddon, K. R. Ionic liquids for clean technology. *J. Chem. Technol. Biotechnol.* 1997, 68, 351–356.
5. Welton, T. Room-temperature ionic liquids—solvents for synthesis and catalysis. *Chem. Rev.* 1999, 99, 2071–2083.
6. Zhao, H., Malhotra, S. V. Applications of ionic liquids in organic synthesis. *Aldrichim. Acta* 2002, 35, 75–83.
7. Earle, M., Forestier, A., Olivier-Bourbigou, H., Wasserscheid, P. In *Ionic Liquids in Synthesis*, ed. P. Wasserscheid, T. Welton, 174–288. Wiley-VCH, Weinheim, 2003.
8. Jain, N., Kumar, A., Chauhan, S., Chauhan, S. M. S. Chemical and biochemical transformations in ionic liquids. *Tetrahedron* 2005, 61, 1015–1060.

9. Wasserscheid, W., Keim, W. Ionic liquids—new "solutions" for transition metal catalysis. *Angew. Chem. Int. Ed.* 2000, 39, 3772–3789.
10. Endres, F., Welton, T. ed. P. Wasserscheid, T. Welton. Inorganic synthesis. In *Ionic Liquids in Synthesis*, ed. P. Wasserscheid, T. Welton, 289–318. Wiley-VCH, Weinheim, 2003.
11. Husum, T. L., Jorgensen, C. T., Christensen, M. W., Kirk, O. Enzyme catalysed synthesis in ambient temperature ionic liquids. *Biocatalysis Biotransform.* 2001, 19, 331–338.
12. Kragl, U., Eckstein, M., Kaftzik, N. Enzyme catalysis in ionic liquids. *Curr. Opin. Biotechnol.* 2002, 13, 565–571.
13. Park, S., Kazlauskas, R. J. Biocatalysis in ionic liquids—advantages beyond green technology. *Curr. Opin. Biotechnol.* 2003, 14, 432–437.
14. Sheldon, R. A., Lau, R. M., Sorgedrager, M. J., van Rantwijk, F., Seddon, K. R. Biocatalysis in ionic liquids. *Green Chem.* 2002, 4, 147–151.
15. van Rantwijk, F., Madeira Lau, R., Sheldon, R. A. Biocatalytic transformations in ionic liquids. *Trends Biotechnol.* 2003, 21, 131–138.
16. van Rantwijk, F., Sheldon, R. A. Biocatalysis in ionic liquids. *Chem. Rev.* 2007, 107, 2757–2785.
17. Kubisa, P. Application of ionic liquids as solvents for polymerization processes. *Prog. Polymer Sci.* 2004, 29, 3–12.
18. Carmichael, A. J., Haddleton, D. M. Polymer synthesis in ionic liquids. In *Ionic Liquids in Synthesis*, ed. P. Wasserscheid, T. Welton, 319–335. Wiley-VCH, Weinheim, 2003.
19. Brennecke, J. F., Maginn, E. J. Purification of gas with liquid ionic compounds. US 6,579,343, 2003.
20. Zhao, H., Xia, S., Ma, P. Use of ionic liquids as "green" solvents for extractions. *J. Chem. Technol. Biotechnol.* 2005, 80, 1089–1096.
21. Zhao, H. Innovative applications of ionic liquids as "green" engineering liquids. *Chem. Eng. Commun.* 2006, 193, 1660–1677.
22. Moniruzzaman, M., Goto, M. Ionic liquids: Future solvents and reagents for pharmaceuticals. *J. Chem. Eng. Jpn.* 2011, 44, 370–381.
23. Siodmiak, T., Marszall, M. P., Proszowska, A. Ionic Liquids: A new strategy in pharmaceutical synthesis. *Mini-Rev. Org. Chem.* 2012, 9, 203–208.
24. Liu, G., Zhong, R., Hu, R., Zhang, F. Applications of ionic liquids in biomedicine. *Biophys. Rev. Lett.* 2012, 7, 121–134.
25. Monteiro, A. L., Zinn, F. K., de Souza, R. F., Dupont, J. Asymmetric hydrogenation of 2-arylacrylic acids catalyzed by immobilized Ru-BINAP complex in 1-n-butyl-3-methylimidazolium tetrafluoroborate molten salt. *Tetrahedron Asymm.* 1997, 8, 177–179.
26. D'Ambra, T. E., Estep, K. G., Bell, M. R., Eissenstat, M. A., Josef, K. A., Ward, S. J., Haycock, D. A., Baizman, E. R., Casiano, F. M. Conformationally restrained analogs of Pravadoline: nanomolar potent, enantioselective, (aminoalkyl)indole agonists of the cannabinoid receptor. *J. Med. Chem.* 1992, 35, 124–135.
27. Earle, M. J., McCormac, P. B., Seddon, K. R. The first high yield green route to a pharmaceutical in a room temperature ionic liquid. *Green Chem.* 2000, 2, 261–262.
28. Yu, H., Wu, J., Ching, C. B. Kinetic resolution of ibuprofen catalyzed by *Candida rugosa* lipase in ionic liquids. *Chirality* 2005, 17, 16–21.

29. Contesini, F. J., de Oliveira Carvalho, P. Esterification of (*RS*)-Ibuprofen by native and commercial lipases in a two-phase system containing ionic liquids. *Tetrahedron Asymm.* 2006, 17, 2069–2073.
30. Scovazzo, P., Visser, A. E., Davis, J. H. J., Rogers, R. D., Koval, C. A., DuBois, D. L., Noble, R. D. Supported ionic liquid membranes and facilitated ionic liquid membranes. In *Ionic Liquids: Industrial Applications for Green Chemistry*, ed. R. D. Rogers, K. R. Seddon, 69–87. Oxford University Press, Washington, DC, 2002.
31. Miyako, E., Maruyama, T., Kamiya, N., Goto, M. Enzyme-facilitated enantioselective transport of (*S*)-ibuprofen through a supported liquid membrane based on ionic liquids. *Chem. Commun.* 2003, 2926–2927.
32. Huh, Y. S., Jun, Y.-S., Hong, Y.-K., Hong, W.-H., Kim, D. H. Microfluidic separation of (*S*)-ibuprofen using enzymatic reaction. *J. Mol. Catal. B Enzym.* 2006, 43, 96–101.
33. Naik, P. U., Nara, S. J., Harjani, J. R., Salunkhe, M. M. Ionic liquid anchored substrate for enzyme catalysed kinetic resolution. *J. Mol. Catal. B Enzym.* 2007, 44, 93–98.
34. Zaidlewicz, M., Sokól, W., Wolan, A., Cytarska, J., Tafelska-Kaczmarek, A., Dzielendziak, A., Prewysz-Kwinto, A. Enolboration of conjugated ketones and synthesis of beta-amino alcohols and boronated alpha-amino acids. *Pure Appl. Chem.* 2003, 75, 1349–1355.
35. Kurata, A., Kitamura, Y., Irie, S., Takemoto, S., Akai, Y., Hirota, Y., Fujita, T., Iwai, K., Furusawa, M., Kishimoto, N. Enzymatic synthesis of caffeic acid phenethyl ester analogues in ionic liquid. *J. Biotechnol.* 2010, 148, 133–138.
36. Uzagare, M. C., Sanghvi, Y. S., Salunkhe, M. M. Application of ionic liquid 1-methoxyethyl-3-methyl imidazolium methanesulfonate in nucleoside chemistry. *Green Chem.* 2003, 5, 370–372.
37. Liu, B. K., Wang, N., Chen, Z. C., Wu, Q., Lin, X. F. Markedly enhancing lipase-catalyzed synthesis of nucleoside drugs' ester by using a mixture system containing organic solvents and ionic liquid. *Bioorg. Med. Chem. Lett.* 2006, 16, 3769–3771.
38. Kumar, V., Malhotra, S. V. Ionic liquid mediated synthesis of 5-halouracil nucleosides: Key precursors for potential antiviral drugs. *Nucleosides Nucleotides Nucleic Acids* 2009 28, 821–834.
39. Shaabani, A., Soleimani, E., Maleki, A. Ionic liquid promoted one-pot synthesis of 3-aminoimidazo[1,2-a]pyridines. *Tetrahedron Lett.* 2006, 47, 3031–3034.
40. Zhang, X., Li, X., Li, D., Qu, G., Wang, J., Loiseau, P. M., Fan, X. Ionic liquid mediated and promoted eco-friendly preparation of thiazolidinone and pyrimidine nucleoside–thiazolidinone hybrids and their antiparasitic activities. *Bioorg. Med. Chem. Lett.* 2009, 19, 6280–6283.
41. Singh, M., Singh, R. S., Banerjee, U. C. Stereoselective synthesis of (*R*)-1-chloro-3(3,4-difluorophenoxy)-2-propanol using lipases from Pseudomonas aeruginosa in ionic liquid-containing system.*J. Mol. Catal. B Enzym.* 2009, 56, 294–299.
42. An, J., Wang, D., Luo, Q., Yuan, X. Antimicrobial active silver nanoparticles and silver/polystyrene core-shell nanoparticles prepared in room-temperature ionic liquid. *Mater. Sci. Eng. C* 2009, 29, 1984–1989.

43. Fan, X., Feng, D., Qu, Y., Zhang, X., Wang, J., Loiseau, P. M., Andrei, G., Snoeck, R., De Clercq, E. Practical and efficient synthesis of pyrano[3,2-c] pyridone, pyrano[4,3-b]pyran and their hybrids with nucleoside as potential antiviral and antileishmanial agents. *Bioorg. Med. Chem. Lett.* 2010, 20, 809–813.
44. Choi, H. J., Uhm, K.-N., Kim, H.-K. Production of chiral compound using recombinant *Escherichia coli* cells co-expressing reductase and glucose dehydrogenase in an ionic liquid/water two phase system. *J. Mol. Catal. B Enzym.* 2011, 70, 114–118.
45. Talisman, I. J., Kumar, V., Deschamps, J. R., Frisch, M., Malhotra, S. V. Application of silver *N*-heterocyclic carbene complexes in *O*-glycosidation reactions. *Carbohydr. Res.* 2011, 346, 2337–2341.
46. Dandia, A., Jain, A. K. Ionic liquid-mediated facile synthesis of novel spiroheterobicyclic rings as potential antifungal and antibacterial drugs. *J. Heterocyclic Chem.* 2013, 50, 104–113.
47. Moniruzzaman, M., Tahara, Y., Tamura, M., Kamiya, N., Goto, M. Ionic liquid-assisted transdermal delivery of sparingly soluble drugs. *Chem. Commun.* 2010, 46, 1452–1454.
48. Mizuuchi, H., Jaitely, V., Murdan, S., Florence, A. T. Room temperature ionic liquids and their mixtures: Potential pharmaceutical solvents. *Eur. J. Pharm. Sci.* 2008, 33, 326–331.
49. Viswanathan, G., Murugesan, S., Pushparaj, V., Nalamasu, O., Ajayan, P. M., Linhardt, R. J. Preparation of biopolymer fibers by electrospinning from room temperature ionic liquids. *Biomacromolecules* 2006, 7, 415–418.
50. Park, T.-J., Lee, S.-H., Simmons, T. J., Martin, J. G., Mousa, S. A., Snezhkova, E. A., Sarnatskaya, V. V., Nikolaev, V. G., Linhardt, R. J. Heparin–cellulose–charcoal composites for drug detoxification prepared using room temperature ionic liquids. *Chem. Commun.* 2008, 5022–5024.
51. Tsioptsias, C., Panayiotou, C. Preparation of cellulose-nanohydroxyapatite composite scaffolds from ionic liquid solutions. *Carbohydr. Polym.* 2008, 74, 99–105.
52. Viau, L., Tourné-Péteilh, C., Devoisselle, J.-M., Vioux, A. Ionogels as drug delivery system: one-step sol–gel synthesis using imidazolium ibuprofenate ionic liquid. *Chem. Commun.* 2010, 46, 228–230.
53. Pang, J., Luan, Y., Li, F., Cai, X., Li, Z. Ionic liquid-assisted synthesis of silica particles and their application in drug release. *Mater. Lett.* 2010, 64, 2509–2512.
54. Mahkam, M., Hosseinzadeh, F., Galehassadi, M. Preparation of ionic liquid functionalized silica nanoparticles for oral drug delivery. *J. Biomater. Nanobiotechnol.* 2012, 3, 391–395.
55. Viçosa, A., Letourneau, J.-J., Espitalier, F., Inês Ré, M. An innovative antisolvent precipitation process as a promising technique to prepare ultrafine rifampicin particles. *J. Crystal Growth* 2012, 342, 80–87.
56. Tsai, T.-H., Thiagarajan, S., Chen, S.-M., Cheng, C.-Y. Ionic liquid assisted synthesis of nano Pd–Au particles and application for the detection of epinephrine, dopamine and uric acid. *Thin Solid Films* 2012, 520, 3054–3059.
57. Moniruzzaman, M., Kamiya, N., Goto, M. Ionic liquid based microemulsion with pharmaceutically accepted components: Formulation and potential applications. *J. Colloid Interface Sci.* 2010, 352, 136–142.

58. Moniruzzaman, M., Tamura, M., Tahara, Y., Kamiya, N., Goto, M. Ionic liquid-in-oil microemulsion as a potential carrier of sparingly soluble drug: Characterization and cytotoxicity evaluation. *Int. J. Pharm.* 2010, 400, 243–250.
59. Dobler, D., Schmidts, T., Klingenhöfer, I., Runkel, F. Ionic liquids as ingredients in topical drug delivery systems.*Int. J. Pharm.* 2013, 441, 620–627.
60. Wermuth, C. G., and Stahl, P. H. In *Handbook of Pharmaceutical Salts: Properties, Selection, and Use*, eds. C. G. Wermuth, P. H. Stahl, 1. Wiley-VCH: Weinheim, Germany, 2002.
61. Kumar, V., and Malhotra, S.V. Ionic liquids as pharmaceutical salts: A historical perspective. In *Ionic Liquid Applications: Pharmaceuticals, Therapeutics, and Biotechnology*, ed. S. V. Malhotra, 1–12. New York: Oxford Press, 2010.
62. Hough, W. L., and Rogers, R. D. Ionic liquids then and now: From solvents to materials to active pharmaceutical ingredients. *Bull. Chem. Soc. Jpn.* 2007, 80, 2262–2269.
63. Cojocaru, O. A., Bica, K., Gurau, G., Narita, A., McCrary, P. D., Shamshina, J. L., Barber, P. S., and Rogers, R.D. Producing ionic liquids: Functionalizing neutral active pharmaceutical ingredients to take advantage of the ionic liquid form. *MedChemComm.* 2013, 4(3), 559–563.
64. Dean, P. M., Turanjanin, J., Yoshizawa-Fujita, M., MacFarlane, D. R., Scott, J. L. Exploring an anti-crystal engineering approach to the preparation of pharmaceutically active ionic liquids. *Cryst. Growth Des.* 2009, 9, 1137–1145.
65. Baglin, I., Mitaine-Offer, A.-C., Nour, M., Tan, K., Cave, C., Lacaille-Dubois, M.-A. A review of natural and modified betulinic, ursolic and echinocystic acid derivatives as potential antitumor and anti-HIV agents. *Mini Rev. Med. Chem.* 2003, 3, 525–539.
66. Cichewicz, R. H., Kouzi, S. A. Chemistry, biological activity, and chemotherapeutic potential of betulinic acid for the prevention and treatment of cancer and HIV infection. *Med. Res. Rev.* 2004, 24, 90–114.
67. Eiznhamer, D. A., Xu, Z. Q. Betulinic acid: a promising anticancer candidate. *IDrugs* 2004, 7, 359–373.
68. Yogeeswari, P., Sriram, D. Betulinic acid and its derivatives: A review on their biological properties. *Curr. Med. Chem.* 2005, 12, 657–666.
69. Krasutsky, P. A. Birch bark research and development. *Nat. Prod. Rep.* 2006, 23, 919–942.
70. Mullauer, F. B., Kessler, J. H., Medema, J. P. Betulinic acid, a natural compound with potent anticancer effects. *Anti-Cancer Drugs* 2010, 21, 215–227.
71. Jäger, S., Winkler, K., Pfüller, U., Scheffler, A. Solubility studies of oleanolic acid and betulinic acid in aqueous solutions and plant extracts of *Viscum album* L. *Planta Med. Planta Med.* 2007, 73, 157–162.
72. Zhao, H., Holmes, S. S., Baker, G. A., Challa, S., Bose, H. S., Song, Z. *J.* Ionic derivatives of betulinic acid as novel HIV-1 protease inhibitors. *Enzyme Inhib. Med. Chem.* 2012, 27, 715–721.
73. Suresh, C., Zhao, H., Gumbs, A., Chetty, C. S., Bose, H. S. New ionic derivatives of betulinic acid as highly potent anti-cancer agents. *Bioorg. Med. Chem. Lett.* 2012, 22, 1734–1738.
74. Ferraz, R., Branco, L. C., Marrucho, I. M., Araújo, J. M. M., Rebelo, L. P. N., da Ponte, M. N., Prudêncio, C., Noronha, J. P., Petrovski, Ž. Development of novel ionic liquids based on ampicillin. *Med. Chem. Commun.* 2012, 494–497.

75. Liu, F., Jiang, Y. Room temperature ionic liquid as matrix medium for the determination of residual solvents in pharmaceuticals by static headspace gas chromatography. *J. Chromatogr. A* 2007, 1167, 116–119.
76. Zeeb, M., Sadeghi, M. Sensitive determination of terazosin in pharmaceutical formulations and biological samples by ionic-liquid microextraction prior to spectrofluorimetry. *Int. J. Anal. Chem.* 2012, article 546282.
77. Bratkowska, D., Fontanals, N., Ronka, S., Trochimczuk, A. W., Borrull, F., Marcé, R. M. Comparison of different imidazolium supported ionic liquid polymeric phases with strong anion-exchange character for the extraction of acidic pharmaceuticals from complex environmental samples. *J. Sep. Sci.* 2012, 35, 1953–1958.

chapter six

Ionic liquids as versatile media for chemical reactions

Jackson D. Scholten, Brenno A. D. Neto, Paulo A. Z. Suarez, and Jairton Dupont

Contents

6.1 Introduction

The need for sustainable and eco-friendly synthetic methodologies is a subject of paramount importance in the modern world. In this sense, the search for processes with both 100% yields and selectivities is a basic requirement to shift the actual paradigm toward ideal chemical transformations. As one can easily imagine, catalytic methodologies naturally hold the master key in this transition. Among all kinds of catalysts, transition metals play a role in the development of modern catalysis.[1–3] Considering the importance of synthetic methodologies for research in chemistry, biology, materials, medicine, and the so-called translational science, it is more than reasonable to affirm that the wedlock of transition metal catalysts with synthetic methodologies is a pathway toward environmental acceptability. Indeed, it is an important part of science itself, with a straight industrial impact. Currently, molecules of almost any complexity may be constructed using catalytic methods. Despite the high level of sophistication reached by transition metal catalysis, many drawbacks must be overcome in this area. For instance, higher values for turnover number (TON) or turnover frequency (TOF) are highly desirable for new catalysts.[4,5] For all of the aforementioned reasons, the search for new reaction conditions, media, and ligands has been observed, aiming for better catalysts and conditions with which to obtain any product.

Among all of the options to improve the catalytic activity of transition metal catalysts, in the last decades, especially since the 1990s, ionic liquids (ILs) have become an exceptional media to test traditional and new catalysts.[6–9] Those liquids may be defined as structures that are entirely composed of ions that melt below 100°C, typically displaying an organic cation associated with a relatively weak coordinating anion. The positive effect of those fluids over a reaction is a consequence of their inherent ionic nature and due to their unique physicochemical properties. Indeed, the origin of this effect and how it operates is the subject of many controversies. The positive effect of ILs on catalytic rates[10] has been well documented over the years. The so-called tunable properties of ILs provide these materials with almost unlimited possibilities, especially when considering the estimation of 10^6 possible combinations of known cations and anions to form an IL.[11] Imidazolium-based ILs constitute an important, rich, and controversial class of ILs, but this has a prominent position among all types of ILs.[12–15] Their physicochemical properties are the subject of major debates, but no one has questioned their importance and applicability. Their negligible vapor pressure, for instance, allows their use for the investigation of the size and shape of metal nanoparticles (NPs) by in situ X-ray photoelectron spectroscopy (XPS)[16] or transmission electron microscopy (TEM)[17] analyses. The supramolecular three-dimensional structural organization of imidazolium-based ILs has a direct impact on their physicochemical properties and helps to explain their influence over several organic transformations.[18] It is not rare to find an example where the reaction yields, rates, and selectivities are improved when a reaction is carried out in ILs.[19–21] ILs are capable of forming ion pairs with charged intermediates and of participating in polar transition states, therefore stabilizing them. Indeed, ILs are able to form large supramolecular aggregates, thus reducing activation barrier energies and facilitating the transformation. This specific property gives ILs the title "entropic drivers," and the aforementioned description is the IL effect. This effect is a consequence of their inherent ionic nature associated with their supramolecular organization. The forces responsible for the high supramolecular organization and nanoorganization of inclusion compounds are the subject of intense debates.[22] For some, H-bonds play a crucial role,[13] whereas others consider the relevance of such bonds minor, attributing high importance to the Coulombic interactions.[23] Another important property to be considered is their capacity to solubilize a great number of organic and inorganic compounds independently of their polarities. This unique property, especially observed in imidazolium-based ILs, is a consequence of their natural segregation in two main domains: polar and nonpolar.[24] As a consequence, polar substrates are preferentially dissolved in polar domains, and vice versa.[25–29] This is of particular importance for the facilitation of close contact among two different reagents when dissolved in the same phase. For all of the desirable characteristics found in imidazolium ILs, this cation was later naturally incorporated as a charge tag in many different ligands (task-specific ionic liquids (TSILs)), which were expected to display similar physicochemical properties to those observed in pure ILs, as reviewed elsewhere.[12,30–32] This elegant strategy enabled mass spectrometry mechanistic evaluations and a better support of novel catalysts in imidazolium ILs. Indeed, as will be discussed, this efficient strategy resulted, overall, in very impressive improvements on yields and selectivities for several reactions. Despite all of the controversies, improved yields, rates, and selectivities are observed, therefore justifying deeper investigation and investment in this field of research. Considering that the number of research groups working in this field has multiplied, we have selected works focusing on hydrogenation, cross-coupling, oxidation, multicomponent reactions, and biomass transformation performed in imidazolium-based ILs using transition metal catalysts.

6.2 *Hydrogenation reactions*

Despite the low solubility of H_2 in ILs,[33] hydrogenation reactions are usually carried out in ILs, most because of the high diffusion degree observed for molecular hydrogen in the ionic phase, thus lessening this drawback. Moreover, the homogeneous nature of the catalyst has been mostly assumed for hydrogenation reactions in ILs.[34] In the 2000s, however, the possibilities of NPs or colloidal systems were discussed.[34,35] In this context, the number of reports describing the synthesis and catalytic applications of transition metal NPs in ILs has increased considerably to date. It is well known that metal NPs can be prepared using both the *top-down* and *bottom-up* approaches, in which the first consists of a physical process where a large particle or bulk metal is fragmented into small units, and the second is based on a chemical reduction/decomposition of metallic precursors.[36,37] The reduction of metal precursors dispersed in ILs in the presence of reducing agents is undoubtedly the most common method applied for the synthesis of metal NPs,[38–53] although studies using physical approaches such as sputtering[54–60] and laser ablation[61–63] have gained more attention in the last years.

It is expected that the observed size, shape, and catalytic activities of metal NPs in ILs are dependent on the nature of the metal precursor, type of IL, and the reaction conditions employed.[24,64] Considering the nature of metal precursors in imidazolium-based ILs, evidence indicates that neutral precursors are located in the nonpolar domains of the IL; for this reason, the size of metal NPs depends on the alkyl side chain length that is attached to the cation.[65–68] However, ionic metal precursors are

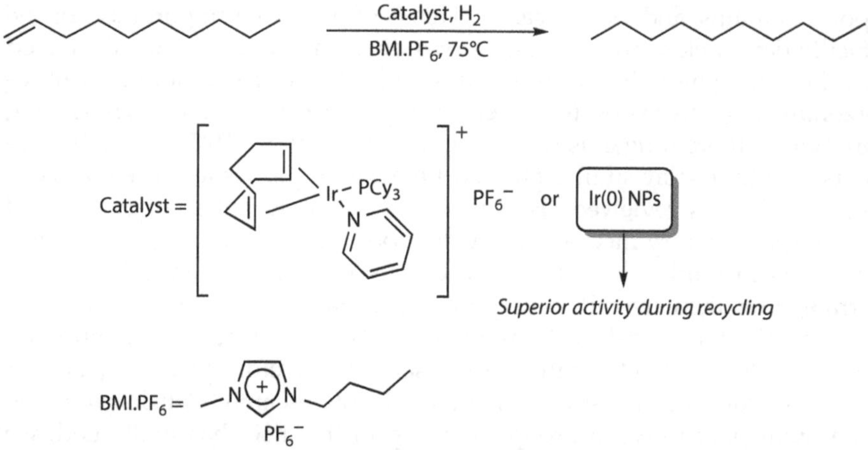

Scheme 6.1 Hydrogenation of 1-decene catalyzed by Crabtree's catalyst and iridium NPs in IL. (From Dupont, J. et al., *J. Am. Chem. Soc.*, 124, 2002, 4228–4229.)

preferentially included at the polar domains, and the NPs size can then be related to the volume of the IL anion.[53,69] For NP preparation in ILs, the use of metal precursors and reducing agents that lessen by-product formation is recommended, because these undesirable species may coordinate over the NP surface (poisoning effect), thereby preventing catalyst activity.[37] Nevertheless, it should be taken into account that the metal NP surface prepared in ILs is never clean, since hydrides, oxides, residual ligands from metal precursor, and carbenes (in the case of imidazolium ILs) are species that are usually present in the medium. Therefore, the activity and selectivity of NPs in hydrogenation reactions may be influenced by the presence of such species, and these issues must be considered for reactions carried out in ILs.[64] One of the first investigations on the hydrogenation of alkenes promoted by metal NPs in ILs was reported at the beginning of the last decade.[38] The biphasic hydrogenation of 1-decene in BMI.PF_6 (BMI = 1-*n*-butyl-3-methylimidazolium) could be performed in the presence of Ir(0) NPs, and these particles were demonstrated to be more active during recycling than the homogeneous Crabtree's catalyst [Ir(COD)py(PCy_3)]PF_6 (COD = 1,5-cyclooctadiene) under the same conditions (Scheme 6.1). Further, the synthetic protocol was extended to the preparation of Pd(0),[42] Rh(0),[39] Ru(0),[70] and Pt(0)[17] NPs in several ILs, and their applications as effective catalysts in hydrogenation reactions were described.

The decomposition of metal carbonyl precursors in ILs also generates effective nanocatalysts for alkene hydrogenation reactions. Rhodium, iridium, and ruthenium metal NPs dispersed in BMI.BF_4 were shown to be active catalysts for cyclohexene hydrogenation.[71,72] It is assumed that metal NPs in ILs have typical homogeneous-like (single-site) catalyst behavior for alkene hydrogenation reactions. In fact, this argument is supported by the observed dependence of the rate of hydrogenation on the structure of the substituted alkenes, i.e., the rate constants decrease in the order mono- > di- > tri- > tetra-substituted double bonds,[73] which is similar to those observed with classical homogeneous catalysts.[74] On the other hand, catalytic insights revealed a typical heterogeneous (multisite) performance of metal NPs for the hydrogenation of arenes.[64] In this context, arene hydrogenation could be employed as a chemical probe for the presence of Ru(0) NPs in IL.[75] Particularly, Ru NPs were formed from the reduction of a Ru(II) precursor by an imidazolium IL in the absence of an additional reducing agent. The hydrogenation of toluene reached 95% conversion, indicating the formation of heterogeneous Ru species, which was further confirmed by transmission electron microscopy (TEM). Indeed, experimental evidence indicates that it is less appropriate to consider homogeneous metal precursors true catalysts during the hydrogenation of arenes.[35,76,77] Moreover, hydrogenolysis products are normally detected during the hydrogenation of substituted arenes and indicate a heterogeneous behavior of NPs. For instance, acetophenone and anisole hydrogenation catalyzed by Ir(0) NPs under solvent-free conditions yield significant amounts of ethylcyclohexane and cyclohexane, respectively, which are derived from the hydrogenolysis of the C-O bond.[39] A hydrogenolysis step was also detected for the hydrogenation of cyclohexenone under the same conditions (Scheme 6.2).[78]

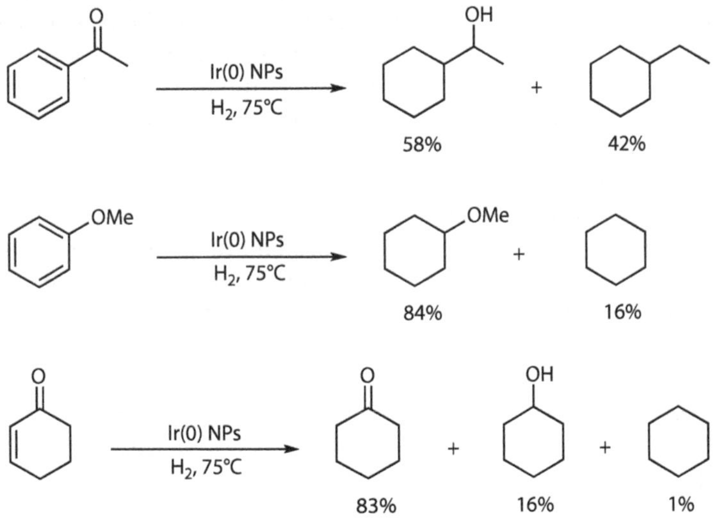

Scheme 6.2 Hydrogenolysis products observed in the hydrogenation of acetophenone, anisole, and cyclohexenone catalyzed by Ir(0) NPs. (From Fonseca, G. S. et al., *Chem. Eur. J.*, 9, 2003, 3263–3269; Fonseca, G. S. et al., *Synlett.*, 2004, 1525–1528.)

The accepted explanation for the formation of hydrogenolysis products is related to the acidity of the metal NP surface. Benzylmethylketone hydrogenation evidenced the heterogeneous behavior of Ir(0) NPs, since considerable selectivity for the saturated ketone (92%) was observed.[78] Although the selective production of cyclohexene from benzene hydrogenation is still a challenge due to the predominant heterogeneous behavior of metal NPs that typically induce total substrate reduction, cyclohexene could be obtained in relative high selectivity at very low benzene conversions using Ru(0) NPs in $BMI.PF_6$.[70] This was possibly due to the lower solubility of the alkene substrate in the IL, which is formed and extracted from the IL phase. Then, the difference in solubility of these compounds can be used to modulate the selectivity of the reactions in ILs.

Interestingly, the hydrogenation of 1,3-cyclohexadiene promoted by Ru(0) NPs in $BMI.NTf_2$ (NTf_2 = *bis*(trifluoromethanesulfonyl)imide) was used as a "structure-sensitive reaction," since the activity and selectivity of the metal NPs were dependent on their size.[79] Here, it was observed that Ru(0) NPs with a size of 2.9 nm were more active than those of 1.1 nm, but the cyclohexene selectivity was higher for the smaller NPs. These results were explained on the basis of the number and type of active sites in each catalyst. While the larger Ru NPs possess a relatively elevated number of adjacent surface sites and exhibit active sites at facial regions, which allow the substrate coordination (superior activity), as well as planar double C=C interactions with full diene hydrogenation, smaller NPs preferably have their active sites located at the vertex and edge positions, thus providing only single π-coordination with a selective formation of the alkene compound.

Active metal NPs can also be prepared using a combination of IL and supercritical carbon dioxide.[80,81] Rh and Pd metal NPs stabilized in an imidazolium IL, as well as in a quaternary ammonium salt, were able to catalyze the selective hydrogenation of acetophenone to 1-phenylethanol, in which Pd(0) NPs in $BMI.PF_6$ were the best recyclable system.[80] In a similar way, Rh(0) NPs were generated in CO_2-induced melting ILs in which, after gas releasing, the NPs were incorporated into the quaternary ammonium salt matrix.[81] In this case, both cyclohexene and benzene could be hydrogenated in the presence of metal NPs (Scheme 6.3).

A general and commonly applied strategy to achieve the enhanced stabilization of metal NPs is the use of functionalized ILs or the addition of extra ligands. In fact, this type of ILs or extra ligands can coordinate at the metal surface, avoiding the aggregation of NPs and, in some cases, inducing selectivity in hydrogenation reactions. To illustrate the applicability of functionalized ILs, the selective hydrogenation of benzonitrile to benzyl-benzylideneamine was performed by Ru(0) NPs dispersed in a nitrile-functionalized IL.[47] Surprisingly, in this system, Ru(0) NPs/IL were not able to catalyze the hydrogenation of toluene. Therefore, it was assumed that the presence of the nitrile group attached to the imidazolium

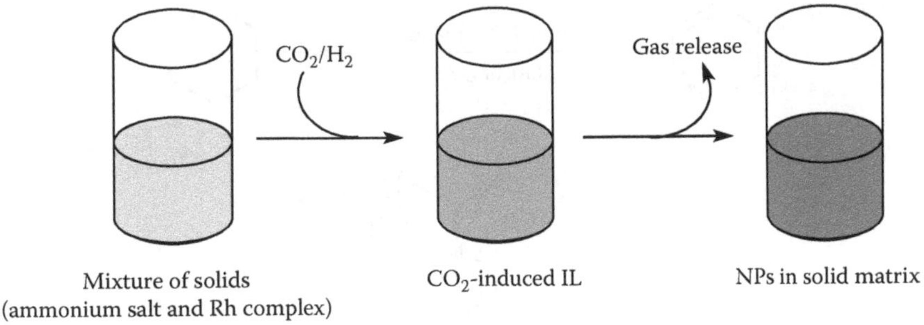

Scheme 6.3 Synthesis of Rh(0) NPs immobilized in a solid matrix by the reduction of $[Rh(acac)(CO)_2]$ in CO_2-induced IL. (Adapted from Cimpeanu, V. et al., *Angew. Chem. Int. Ed.*, 48, 2009, 1085–1088.)

cation can interact with the metal surface and, as a consequence, modulate the selectivity of the NPs depending on the substrate. The same functionalized IL was tested on the hydrogenation of alkynes using Pd(0) NPs.[82] In this case, the selectivity for the alkene or alkane products could be controlled by only adjusting the hydrogen gas pressure. Indeed, the corresponding alkene was selectively produced under a low hydrogen gas pressure (1 bar), while the alkane was observed at a higher pressure (4 bar). The presence of additional stabilizers such as polymers,[83,84] tetraalkylammonium salts,[85] and N-donor ligands[40,86–90] provides a positive effect on the stabilization of metal NPs in ILs. Polymers are extensively used as extra capping ligands due to their ability to weakly interact with the metal surface, providing an efficient steric protection for the NPs. Among the polymers employed, poly(N-vinyl-2-pyrrolidone) (PVP) has been widely studied. However, PVP has a limited solubility in ILs, and for this reason, it is necessary introduce some functionality on its structure or use functionalized ILs in order to dissolve the polymer into the IL phase. In fact, PVP or its derived ionic compounds have been tested as extra stabilizers for Pt(0) NPs in ILs on the selective transformation of chloronitroarenes to the aromatic chloroamines.[83,91] Similarly, these ligands were adopted for the hydrogenation of alkenes and arenes promoted by Rh(0) NPs in BMI.X (X = BF_4, PF_6) or styrene reduction in hydroxyl-functionalized ILs (Scheme 6.4).[84,92–94]

In the case of quaternary ammonium salts, the system composed of Pd(0) NPs stabilized by tetrabutylammonium bromide (TBAB) in BMI.PF_6 was active and selective for the hydrogenation of several unsaturated substrates.[85] As an example for N-donor stabilizers, the influence of bipyridine ligands (2,2′-, 3,3′-, 4,4′-bipyridine) on the activity and selectivity of Rh(0) NPs was evaluated during styrene hydrogenation.[86] The observed activities and selectivities were attributed to the different coordination modes of the bipyridine ligands with the metal NPs, and the best results were

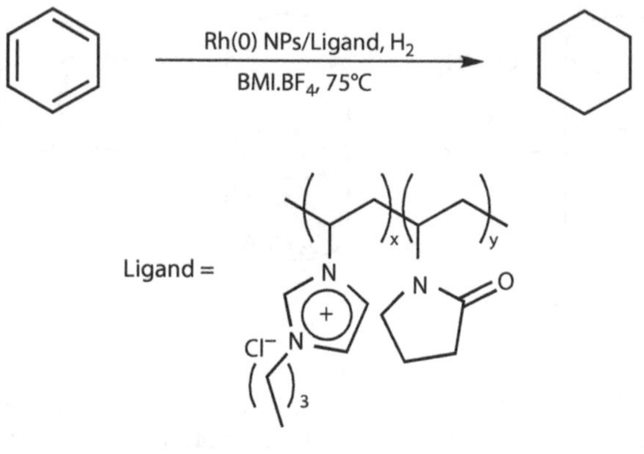

Scheme 6.4 Benzene hydrogenation catalyzed by rhodium NPs in the presence of poly-[(N-vinyl-2-pyrrolidone)-*co*-(1-vinyl-3-butylimidazolium chloride)] copolymer and an imidazolium IL. (From Mu, X. D. et al., *J. Am. Chem. Soc.*, 127, 2005, 9694–9695.)

attained using 3,3′- and 4,4′-bipyridine. These ligands probably interact with the metal surface by a monodentate mode, allowing the approach of the substrate, which is in contrast to 2,2′-bipyridine that blocks the active sites and reduces the access of the reactant. Bipyridine-derived ionophilic ligands were also evaluated as additional capping agents for Rh(0) NPs in $BMI.NTf_2$.[87] The NPs stabilized using the ionic ligand containing a longer alkyl chain between the imidazolium group and the pyridine skeleton showed superior activity in the hydrogenation of substituted arenes compared to the other ligands tested. The activity followed the trend $BIHB.(NTf_2)_2$ > 2,2′-bipy >> $BIMB.(NTf_2)_2$, where BIHB corresponds to 4,4′-bis-[7-(2,3-dimethylimidazolium)heptyl]-2,2′-bipyridine and BIMB is 4,4′-bis[(1,2-dimethylimidazolium)methyl]-2,2′-bipyridine. The poor activity detected for the $BIMB.(NTf_2)_2$ ligand may be related to electron-withdrawing effects, the undesirable minor proximity of positive charges to the NP's surface, and steric hindrance. Other N-donor ligands such as phenanthroline[40] and $BM_2(DPA)I.PF_6$ ($BM_2(DPA)I$ = 2,3-dimethyl-1-[3-N,N-bis(2-pyridyl)-propylamido]imidazolium)[95] have also been used as extra stabilizers for Pd(0) NPs in ILs and tested during alkene hydrogenation.

In order to obtain a suitable and more eco-friendly process, the SCILL (solid catalysts with an ionic liquid layer) method has been suggested as an alternative to exploit the singular properties of ILs and at the same time reduce their amounts in catalytic reactions.[96–98] In this protocol, classical heterogeneous catalysts are modified with a thin film of IL, giving the desired properties for the solid material, such as activity and selectivity. In addition, the SCILL method considerably reduces the mass transfer limitations observed in reactions when IL is used as a solvent. The synthesis of Pd(0) NPs immobilized in a thin film of $BMI.BF_4$ and the use of this system for the hydrogenation of 1,3-butadiene to 1-butene have been demonstrated.[42] Here, the selectivity is a consequence of the higher solubility of the diene, which is at least four times greater in $BMI.BF_4$ than that observed for butenes. Using a similar idea, SILP (supported ionic liquid phase) catalysis has emerged as a potential system that also provides the IL benefits to the solid supports. Although the concepts of the SCILL and SILP methods are slightly different, in some cases they have been used randomly. In fact, while the SCILL protocol modifies the classical supports with a thin film of IL, the SILP method involves the use of IL molecules attached covalently to the support surface.[64] Pd(0) NPs supported on silica gel modified by ILs can promote the selective hydrogenation of cinnamaldehyde to hydrocinnamaldehyde.[99] The hydrogenation of alkynes catalyzed by Pt(0) NPs supported on magnetite NPs modified with an imidazolium IL mainly afforded alkenes as products.[100] Notably, for α,β-unsaturated aldehydes, selective reduction of the carbonyl moiety was observed (Scheme 6.5).

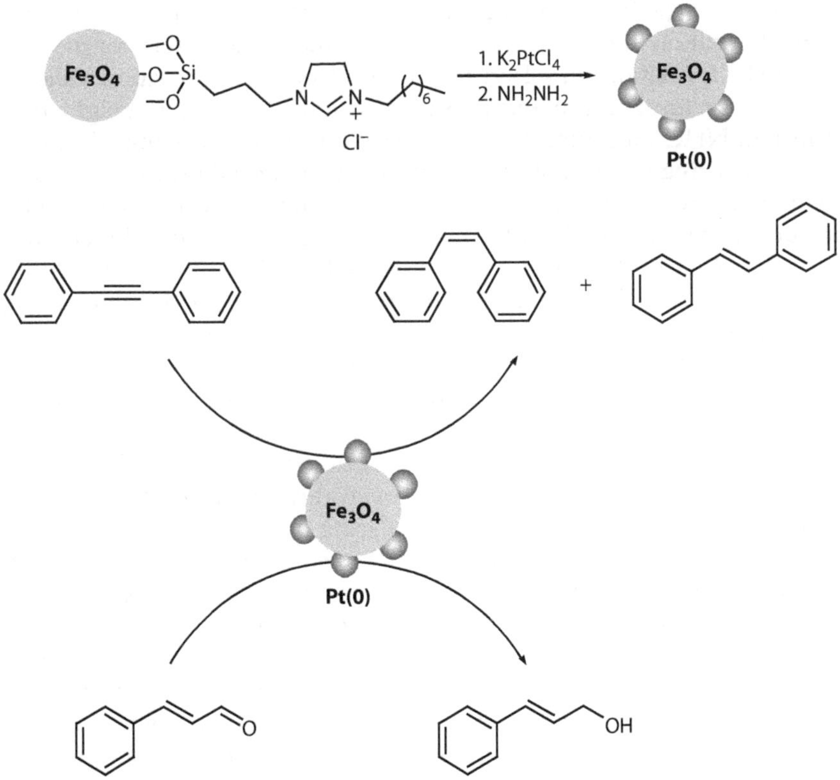

Scheme 6.5 Preparation and catalytic application of magnetite-supported Pt(0) NPs. (Adapted from Abu-Reziq, R. et al., *Adv. Synth. Catal.*, 349, 2007, 2145–2150.)

The in situ deposition of Pd(0) NPs on IL-functionalized multiwalled carbon nanotubes (IL.X.MWCNTs; X is the IL anion) provided an active catalyst for the hydrogenation of *trans*-stilbene to 1,2-diphenylethane.[101] Other developed materials such as Pd(0) NPs supported on molecular sieves functionalized with an IL layer[102] and Rh(0) or Pt(0) NPs immobilized in hybrid membranes containing IL/cellulose acetate[103] can be mentioned. These catalytic systems showed excellent activity during cyclohexene hydrogenation.[102,103] In general, both SCILL and SILP methods produce more robust/recyclable and active catalysts than the conventional biphasic systems, and also reduce the problems associated with mass transfer limitations and low gas solubility that are commonly observed in biphasic media.[64]

6.3 Cross-coupling reactions

The Heck cross-coupling reaction is among the most important for Pd-catalyzed C-C bond formation. The presence of acidic hydrogen at the C2 position of the imidazolium ring allows direct N-heterocyclic carbene (NHC) formation in the presence of a base, which is necessary for the reaction. NHC formation may lead to a large drop in the reaction yields.[104] In this sense, ammonium and phosphonium ILs have, at first sight, been preferred. NHC formation from imidazolium derivatives may, however, be used favoring the formation of a stable and active Pd catalyst, as will be discussed. Pore and coworkers have described the synthesis of a hydrophobic fluorine-containing IL for the ligand-free Heck reaction (Scheme 6.6).[105] The authors tested five commercially available Pd complexes and found that $Pd(OAc)_2$ was the most active when supported in IL **4**. Interestingly, the authors showed that Pd NPs form in situ and with variable size in the range of 20–100 nm, which was confirmed by TEM analysis. IL **4** was acting as the stabilizing agent for the metal NPs, and interestingly, two examples with aryl chlorides were reported. Recycling studies were also performed and a slight loss of activity was noted for the five runs.

Charge-tagged ligands have also been developed to form Pd complexes. An imidazolium-tagged TSIL bearing a triazole moiety has been synthesized and applied as an efficient ligand to stabilize the Pd center applied for Suzuki coupling reactions (Scheme 6.7).[106] Fourteen ligands were synthesized, and one of them ($R^1 = R^2 =$ Me, $R^3 = {}^n$Bu) showed an efficient stabilization of the Pd center. Considering the tunable properties of the imidazolium chemistry, the authors envisaged the reaction in water. The complex of **8** with $PdCl_2$ showed good solubility and was applied for the Suzuki reaction. Aryl bromides and iodides gave the desired product with almost quantitative yields. The novel imidazolium-linked triazole derivatives could also be used as solvents for the reaction with excellent results.

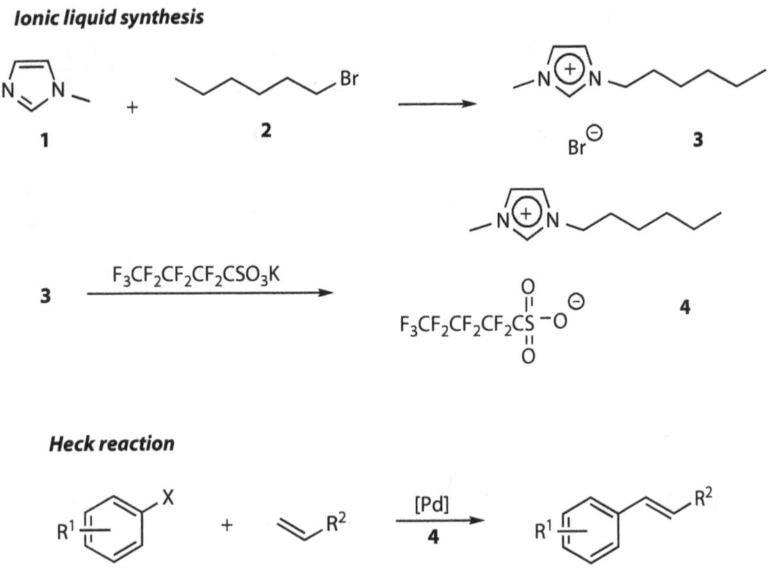

Scheme 6.6 Synthesis and application of a fluorine-containing IL as the medium for the Heck reaction. (From Gaikwad, D. S. et al., *Tetrahedron Lett.*, 53, 2012, 3077–3081.)

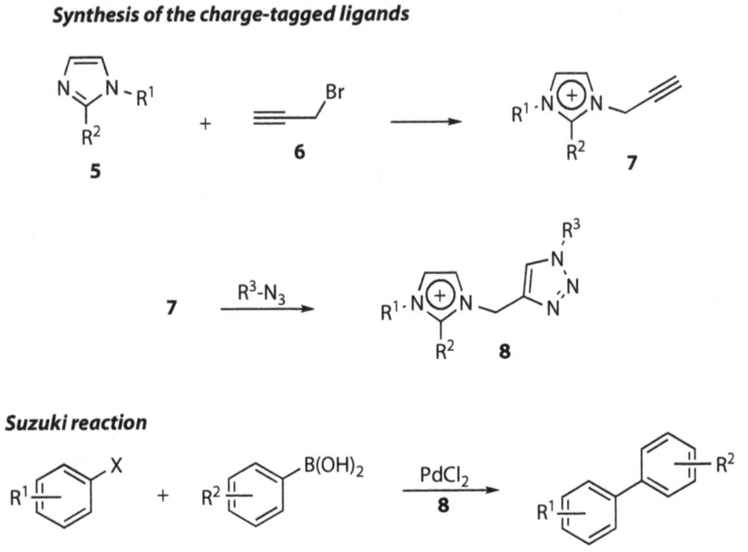

Scheme 6.7 Synthesis and application of charge-tagged ligands for the Suzuki reactions. (From Li, L. Y. et al., *Chem. Eur. J.*, 18, 2012, 7842–7851.)

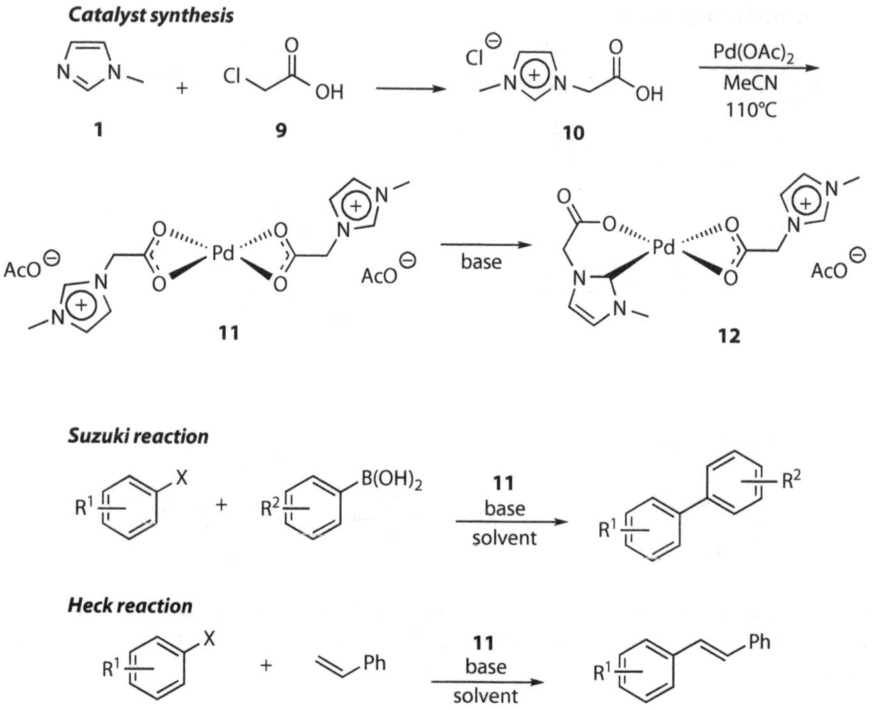

Scheme 6.8 In situ generation of the active charge-tagged organometallic catalyst for the phosphine-free Heck and Suzuki reactions. (From Oliveira, F. F. D. et al., *J. Org. Chem.*, 76, 2011, 10140–10147.)

Another interesting example of the use of charge-tagged ligands has been recently demonstrated.[107] An acetate derivative was tagged with an imidazolium ring to an in situ generation of NHC, which in turn cyclized, forming the active organometallic palladium catalyst applied in Suzuki and Heck phosphine-free reactions (Scheme 6.8). It was demonstrated that the precatalyst **11** is, indeed, ca. 14 times more active than $Pd(OAc)_2$ for the Heck reaction.

6.4 Oxidation reactions

The high potential of ILs and TSILs as reaction media or catalysts for oxidation reactions has not escaped the interest of many research groups, as already reviewed.[108] An interesting oxidation reaction recently described allows the direct transformation of benzyl chloride (bromide or iodide) to benzaldehyde in the presence of EMI.Cl (EMI = 1-*n*-ethyl-3-methylimidazolium) under microwave irradiation (Scheme 6.9).[109] The reaction

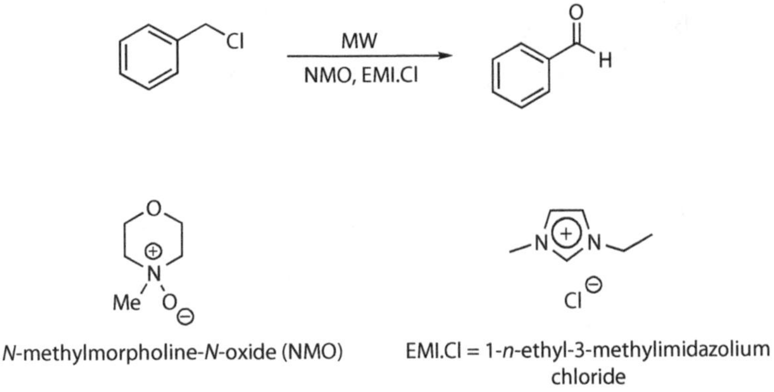

Scheme 6.9 Oxidation reaction of benzyl chloride to benzaldehyde in the presence of EMI.Cl. (From Khumraksa, B. et al., *Tetrahedron Lett.*, 54, 2013, 1983–1986.)

proceeds in only 1 min (150 W, 100°C), and an additive (KI) was proven to have a beneficial effect over the reaction.

When the methodology was applied to other substrates, interesting products could be obtained in high yields, as shown in Scheme 6.10.

TSILs have been successfully used in oxidative desulfurization of fuels (Scheme 6.11).[110] Dupont and coworkers demonstrated that the functionalization of a carboxylic acid derivative with an imidazolium cation

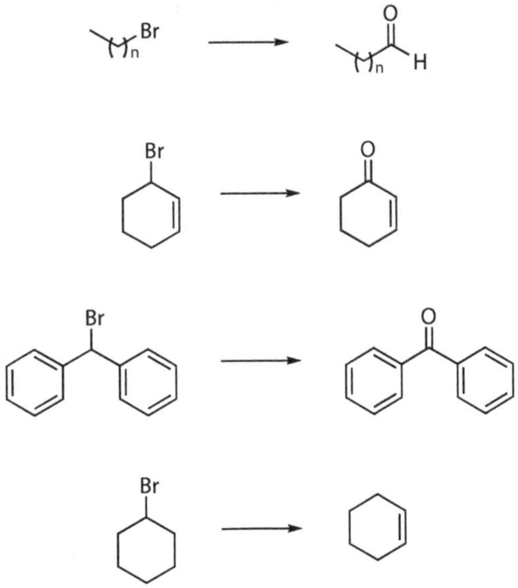

Scheme 6.10 Oxidation reactions under microwave irradiation in the presence of EMI.Cl. (From Khumraksa, B. et al., *Tetrahedron Lett.*, 54, 2013, 1983–1986.)

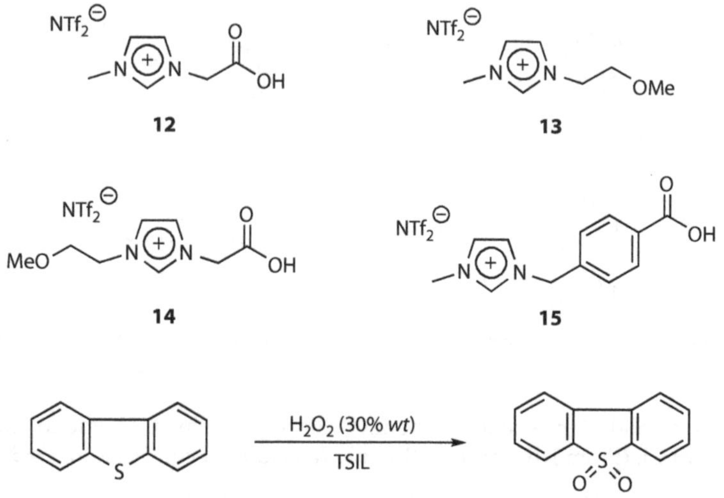

Scheme 6.11 TSILs used for oxidative desulfurization reactions of dibenzothiophene using hydrogen peroxide as the oxidizing agent. The association with the stable and hydrophobic NTf_2^- anion proved to be an important feature. (From Lissner, E. et al., *ChemSusChem*, 2, 2009, 962–964.)

yielded a TSIL that proved to be a successful alternative medium for the oxidative desulfurization reaction using hydrogen peroxide as the oxidizing agent when associated with a highly stable and hydrophobic anion (i.e., NTf_2^-).

6.5 Multicomponent reactions

Multicomponent reactions (MCRs) are elegant and important tools used for the generation of a product in a single step (one pot) from three or more reagents displaying high atom economy and multiple-bond-forming efficiency.[111–113] Some interesting features associate MCRs with ILs. Some have affirmed that the combination of MCRs performed in ILs is "a perfect synergy for eco-compatible heterocyclic synthesis."[114] Also, it has been said that this combination is "a promising strategy for the development of valuable eco-compatible organic synthesis procedures"[114] and "an excellent media for multiple bond-forming transformations."[114] These sentences are not overrated, and indeed, one can observe a huge effort toward the development of new catalysts to perform MCRs in ILs. A recent example is the synthesis of coumarin derivatives (furocoumarins) using the MCR approach (Scheme 6.12).[115] All coumarins **17** were synthesized above the 80% yield. IL **16** was used as the promoter and as the reaction medium for the MCR.

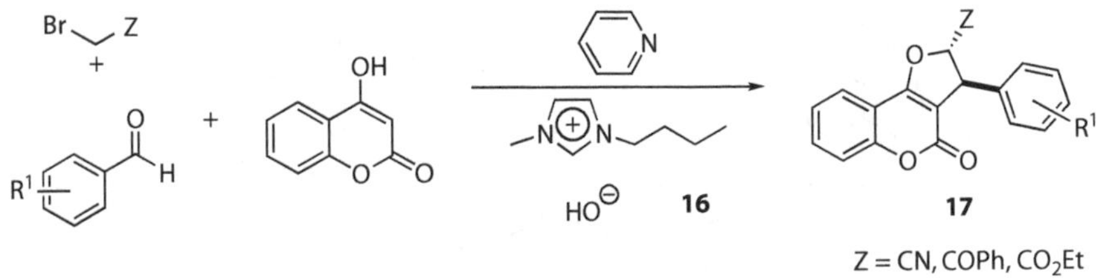

Scheme 6.12 Synthesis of furocoumarins with the MCR approach. Note that the IL **16** is used as the reaction medium and as the promoter of the reaction. (From Rajesh, S. M. et al., *Tetrahedron*, 68, 2012, 5631–5636.)

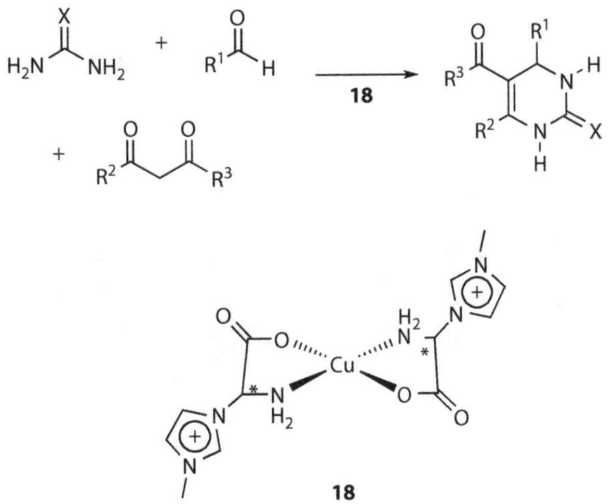

Scheme 6.13 Enantioselective Biginelli reaction with a copper complex with an ionically tagged copper complex with chiral ligands. (From Karthikeyan, P. et al., *Organomet. Chem.*, 723, 2013, 154–162.)

An enantioselective approach for the Biginelli reaction using a novel copper complex with chiral ligands was recently described (Scheme 6.13).[116] The authors observed *ee* above 60% and yields ranging from 37 to 95%. Reactions were conducted at 25°C for 12–24 h. The authors noted that the use of base reduced both yields and *ee*.

The mechanism of the Biginelli reaction is a controversial issue, especially for Lewis acid-catalyzed reactions. In this sense, two recent contributions shed some light on those reactions performed in ILs.[117,118] The authors followed the reaction using NMR and ESI-MS(/MS) to identify key intermediates of the MCR. Scheme 6.14 shows a plausible explanation and the role of the reagents in the formation of key reactive intermediates.

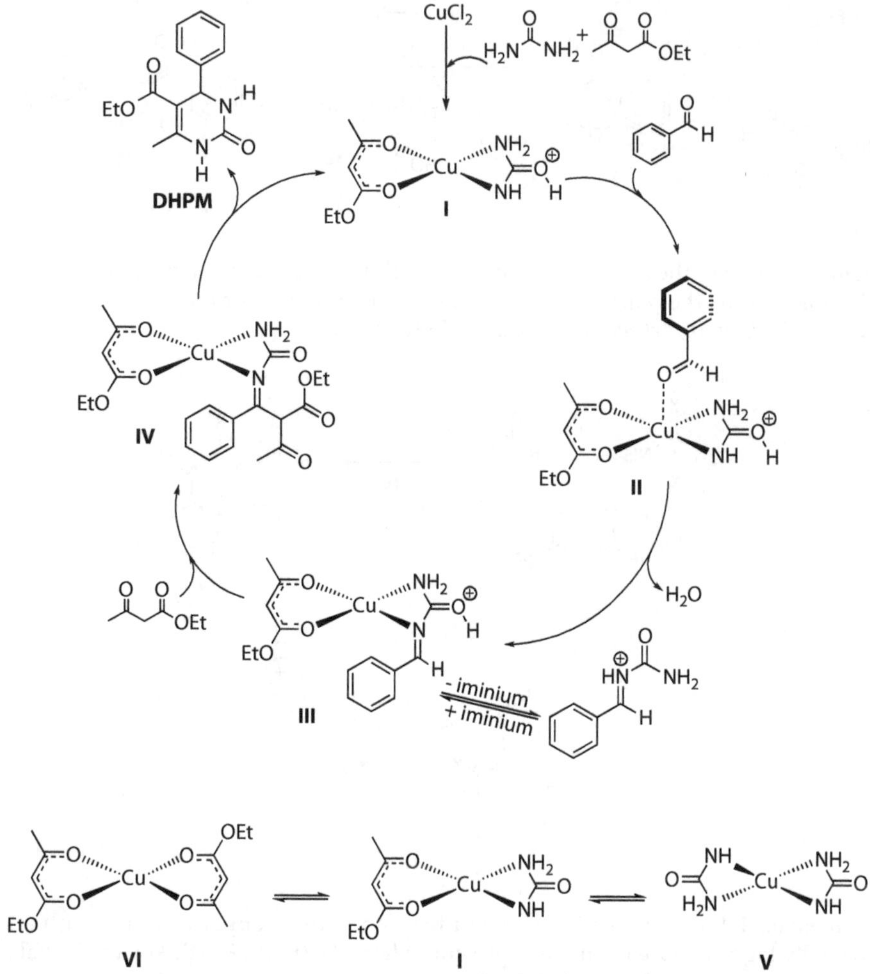

Scheme 6.14 Plausible mechanism for Lewis acid-catalyzed Biginelli reaction in ILs. Scheme shown for a copper-catalyzed reaction. (Adapted from Ramos, L. M. et al., *J. Org. Chem.*, 77, 2012, 10184–10193.)

It has been shown that the PF_6^- anion is responsible for ruling the reagent approximation to form intermediate **II** and for the formation of intermediate **III**. This is a good example of an IL acting as an "entropic driver"[24] for a catalyzed reaction performed in such ionic fluids.

6.6 Biomass transformation

The advantageous physicochemical properties of ILs can also be explored for the transformation of biomass into valuable chemical compounds.[119–123]

Common catalytic methods in ILs have employed acids,[124–130] enzymes,[131,132] metal nanoparticles,[133–135] and metal salts[136–148] as efficient promoters for the conversion of biomass-derived substrates. Among these methods, acid hydrolysis of biomass feedstocks in ILs became a simple and efficient protocol to obtain the desired products, such as 5-hydroxymethyl-2-furfural (HMF), which is an important building block for the synthesis of biofuels and chemicals. In fact, the dehydration of fructose into HMF could be carried out in BMI.Cl/acetone at 25°C, using a strong acidic ion exchange resin as a catalyst.[125] Moreover, the presence of by-products such as glucose, levulinic acid, and formic acid was also verified in small amounts (1–2%). An efficient one-pot procedure to synthesize HMF from inulin in ILs has also been reported.[127] First, the inulin was decomposed to fructose (84% yield) in the presence of BMI.HSO_4, and in a second step, BMI.Cl and an acidic ion exchange resin were introduced to transform the fructose to HMF (82% yield). A two-step approach for the catalytic conversion of glucose to 2,5-dimethylfuran in IL/acetonitrile was developed.[128] This procedure involved the dehydration of glucose to HMF via acid catalysis and the subsequent hydrogenation of HMF to 2,5-dimethylfuran catalyzed by Pd/C.

Metal salts in ILs constitute alternative systems for the production of HMF. When fructose or glucose is dissolved in the EMI.Cl IL/metal chloride system (EMI = 1-ethyl-3-methylimidazolium), satisfactory yields of HMF (63–83% for fructose and 68% for glucose) can be produced at 80–100°C (Scheme 6.15).[136]

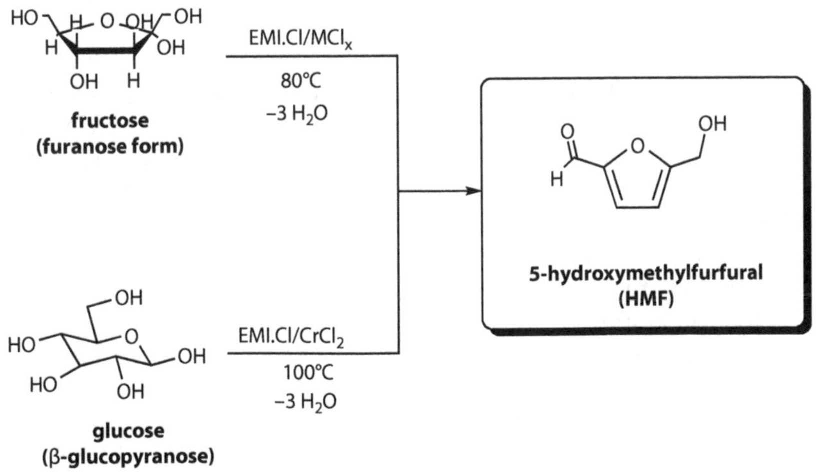

Scheme 6.15 Fructose and glucose dehydration catalyzed by metal chlorides in ILs. (Adapted from Zhao, H. B. et al., *Science*, 316, 2007, 1597–1600.)

In the case of glucose, the activity of the Cr catalyst was related to the presence of $CrCl_3^-$ species that are involved in the mutarotation of glucose, giving a mixture of α- and β-anomers and the subsequent isomerization to fructose. Noteworthy, the isomerization process is a key step to begin a fast dehydration reaction. Similarly, glucose dehydration catalyzed by $SnCl_4$ in $EMI.BF_4$ followed a mechanism through a five-membered ring chelate intermediate, which can be converted into an enediol species and then isomerized to fructose or to directly yield the HMF product.[137] Interestingly, this catalytic system was active for other types of biomass feedstocks: fructose, sucrose, cellobiose, inulin, and starch. Based on density functional theory (DFT) calculations and X-ray absorption spectroscopy (XAS) experiments, the activity of $CrCl_2$ for the glucose dehydration in IL was explained by the formation of $CrCl_4^{2-}$ and the replacement of a chlorine ligand by one hydroxyl group from the substrate.[149,150] The chloride anion of IL also apparently acts as an efficient basic mediator, making interactions with the hydroxyl groups of the carbohydrate by hydrogen bonds, which allows the conversion of glucose into HMF. Although this is in accordance with the finding that the facile isomerization of glucose to fructose promoted by the Cr catalyst in IL is responsible for the high HMF yields, there is no chemical bonding between the metal center and the cationic imidazolium moiety, which is in contrast to earlier assumptions.[136]

A tungsten salt (WCl_6) has also been reported for the dehydration of fructose to HMF in BMI.Cl IL.[138] This system was shown to be competent, even at low temperatures, where up to 63% HMF could be obtained at 50°C; in a biphasic condition using THF/IL, the yield of the product increased to 72%. This positive effect of the biphasic system allows the recyclability of the IL phase, and HMF was produced at large scales by a continuous reaction. A catalytic system containing $GeCl_4$ in ILs was highly effective for the generation of HMF (92%) from fructose, but moderate yields were verified using glucose, cellobiose, and cellulose.[139] Lanthanide-based catalysts were also evaluated under glucose transformation in imidazolium ILs.[140] In this case, ytterbium(III) chloride or triflate in BMI.Cl at 140°C was the most active system, although only moderate yields in HMF could be detected (up to 24%). It was recognized that the acidic character of the C2-H position of the imidazolium ring plays a significant role during fructose dehydration in ILs,[151] which is in agreement with previous results.[136] For this reason, the Brønsted acidity of several ILs was explored for the conversion of biomass-derived substrates.[152–154] Ammonium salts have also been used to promote fructose dehydration.[155] Even at high concentrations of fructose (33 and 50 wt%), tetraethylammonium chloride was able to provide elevated yields of HMF (81 and 79%, respectively) at 120°C. For the reaction employing the higher substrate concentration, $NaHSO_4.H_2O$ was used as a co-catalyst. In addition, the use of $FeCl_3.H_2O/Et_4N.Br$

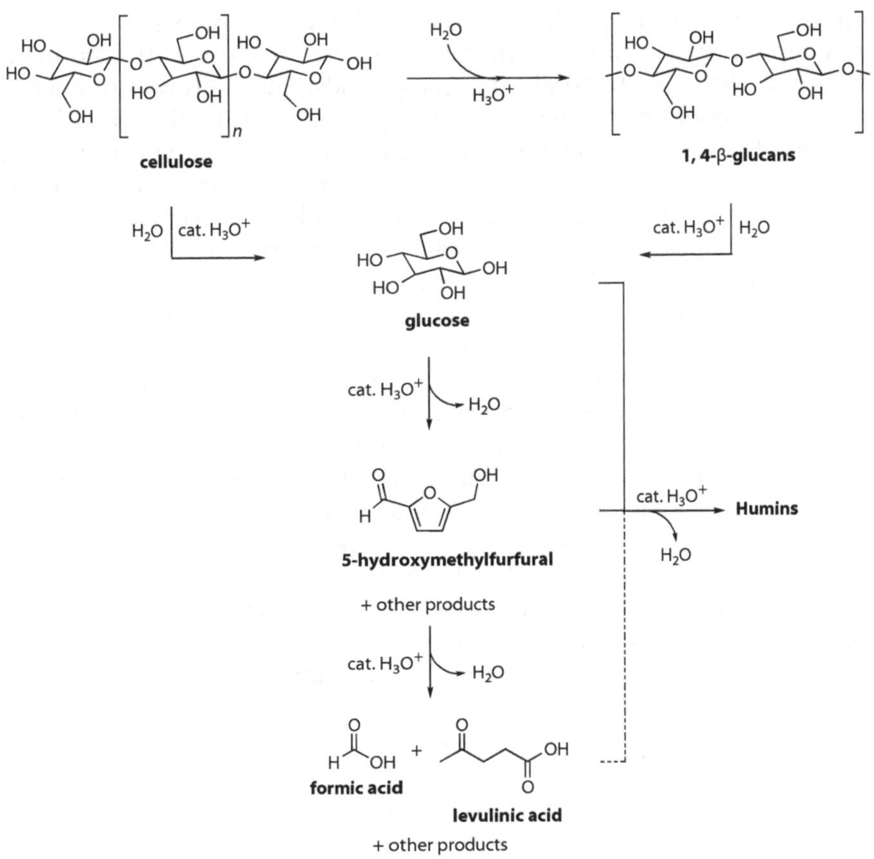

Scheme 6.16 Example of products obtained by acid-catalyzed cellulose depolymerization. (From Rinaldi, R. et al., *ChemSusChem*, 3, 2010, 266–276.)

was found to be an efficient catalytic system for the conversion of sugars (86% HMF yield at 90°C).[141]

Complex biomass feedstocks such as cellulose and lignocellulose are interesting starting materials for the synthesis of fuels and other chemical compounds. In the presence of acid catalysts, various studies revealed the hydrolysis of cellulose dissolved in ILs. For example, solid acids in ILs were described as a suitable medium for the depolymerization of cellulose (Scheme 6.16).[124,129] In addition, factors involving the dehydration of glucose and the formation of humins after cellulose hydrolysis were also evaluated.[130]

In the case of metal salts as promoters, a system based on DMA-LiCl/metal chloride/HCl/EMI.Cl constituted an active medium to dissolve and transform cellulose and lignocellulosic substrates.[142] The conversion of cellulose into HMF (54% yield) was catalyzed by $CrCl_2$ at 140°C, which can be considered mild conditions compared with the typically high temperatures that are required to promote the acid hydrolysis of cellulose. In a similar way, corn stover (lignocellulosic biomass) was transformed into HMF (48%) and furfural (34%) compounds. The hydrolysis of microcrystalline cellulose was performed using a task-specific IL 1-(4-sulfonic acid)butyl-3-methylimidazolium hydrogen sulfate in the presence of $FeCl_2$ and $MnCl_2.4H_2O$.[143,144] Although a small amount of levulinic acid was generated when $FeCl_2$ was the catalyst, similar yields of HMF and furfural products were achieved for both metal catalysts. Moreover, the IL phase was recycled for at least five runs, but a slight decrease in activity could be observed. Attempts to convert cellulose and lignocellulosic materials promoted by a mixture of metal chlorides in IL ($CrCl_2/RuCl_3$,Cr:Ru = 4:1) at 120°C produced good yields of products.[145] In addition, a microwave-assisted method was also tested for the conversion of cellulose and lignocellulosic biomass in ILs.[146,147] In a few minutes of reaction, cellulose was converted to HMF (62% of yield) using $CrCl_3.6H_2O$ dispersed in BMI.Cl, while it was possible to reach yields of 52% of HMF and 31% of furfural from lignocellulosic materials. Under microwave irradiation at 160°C, the reusable catalytic system $CrCl_3$/LiCl in BMI.Cl was competent for cellulose hydrolysis, giving yields of HMF up to 62%, and for untreated wheat straw, a 61% yield of HMF and 43% of furfural was detected.[148] In these cases, the IL was recycled for at least three runs without any loss of catalyst activity.

6.7 Conclusions and remarks

There is no doubt that ILs provide an important liquid platform for the development of multiphase catalytic systems for various transformations. Classical homotopic and heterotopic catalysts are relativity easily immobilized in ILs without—in most cases—losing their catalytic performance. Moreover, ILs provide a medium for the development of new types of catalysts, such as metal nanoparticles, or more interestingly for the solubilization of biomass feedstocks; in particular, this was the case for those abundant biomaterials that are almost impossible to solubilize in other fluids, thus allowing their catalytic transformation in more added-value chemicals. It is therefore expected that ILs will play an even more important role, not only in classical catalytic processes, but also in biomass transformation.

References

1. Lipshutz, B. H., Ghorai, S. Transition-metal-catalyzed cross-couplings going green: In water at room temperature. *Aldrichim. Acta* 2008, 41, 59–72.
2. Liu, S. F., Xiao, J. L. Toward green catalytic synthesis—Transition metal-catalyzed reactions in non-conventional media. *J. Mol. Catal. A Chem.* 2007, 270, 1–43.

3. Yan, N., Xiao, C. X., Kou, Y. Transition metal nanoparticle catalysis in green solvents. *Coord. Chem. Rev.* 2010, 254, 1179–1218.
4. Umpierre, A. P., de Jesus, E., Dupont, J. Turnover numbers and soluble metal nanoparticles. *ChemCatChem* 2011, 3, 1413–1418.
5. Kozuch, S., Martin, J. M. L. "Turning over" definitions in catalytic cycles. *ACS Catal.* 2012, 2, 2787–2794.
6. Dupont, J., de Souza, R. F., Suarez, P. A. Z. Ionic liquid (molten salt) phase organometallic catalysis. *Chem. Rev.* 2002, 102, 3667–3691.
7. Welton, T. Room-temperature ionic liquids. Solvents for synthesis and catalysis. *Chem. Rev.* 1999, 99, 2071–2083.
8. Earle, M. J., Seddon, K. R. Ionic liquids. Green solvents for the future. *Pure Appl. Chem.* 2000, 72, 1391–1398.
9. Welton, T. Ionic liquids in catalysis. *Coord. Chem. Rev.* 2004, 248, 2459–2477.
10. Lee, J. W., Shin, J. Y., Chun, Y. S., Bin Jang, H., Song, C. E., Lee, S. G. Toward understanding the origin of positive effects of ionic liquids on catalysis: Formation of more reactive catalysts and stabilization of reactive intermediates and transition states in ionic liquids. *Acc. Chem. Res.* 2010, 43, 985–994.
11. Weingartner, H., Cabrele, C., Herrmann, C. How ionic liquids can help to stabilize native proteins. Phys. Chem. *Chem. Phys.* 2012, 14, 415–426.
12. Neto, B. A. D., Spencer, J. The impressive chemistry, applications and features of ionic liquids: Properties, catalysis & catalysts and trends. *J. Braz. Chem. Soc.* 2012, 23, 987–1007.
13. Dong, K., Zhang, S. J. Hydrogen bonds: A structural insight into ionic liquids. *Chem. Eur. J.* 2012, 18, 2748–2761.
14. Consorti, C. S., de Souza, R. F., Dupont, J., Suarez, P. A. Z. Dialkylimidazolium cation based ionic liquids: Structure, physico-chemical properties and solution behaviour. *Quim. Nova* 2001, 24, 830–837.
15. Dupont, J., Consorti, C. S., Spencer, J. Room temperature molten salts: Neoteric "green" solvents for chemical reactions and processes. *J. Braz. Chem. Soc.* 2000, 11, 337–344.
16. Bernardi, F., Scholten, J. D., Fecher, G. H., Dupont, J., Morais, J. Probing the chemical interaction between iridium nanoparticles and ionic liquid by XPS analysis. *Chem. Phys. Lett.* 2009, 479, 113–116.
17. Scheeren, C. W., Machado, G., Dupont, J., Fichtner, P. F. P., Texeira, S. R. Nanoscale Pt(0) particles prepared in imidazolium room temperature ionic liquids: Synthesis from an organometallic precursor, characterization, and catalytic properties in hydrogenation reactions. *Inorg. Chem.* 2003, 42, 4738–4742.
18. Dupont, J. On the solid, liquid and solution structural organization of imidazolium ionic liquids. *J. Braz. Chem. Soc.* 2004, 15, 341–350.
19. De Bellefon, C., Pollet, E., Grenouillet, P. Molten salts (ionic liquids) to improve the activity, selectivity and stability of the palladium catalyzed Trost-Tsuji C-C coupling in biphasic media. *J. Mol. Catal. A Chem.* 1999, 145, 121–126.
20. Dubreuil, J. F., Bazureau, J. P. Rate accelerations of 1,3-dipolar cycloaddition reactions in ionic liquids. *Tetrahedron Lett.* 2000, 41, 7351–7355.
21. Chauvin, Y., Mussmann, L., Olivier, H. A novel class of versatile solvents for two-phase catalysis: Hydrogenation, isomerization, and hydroformylation of alkenes catalyzed by rhodium complexes in liquid 1,3-dialkylimidazolium salts. *Angew. Chem. Int. Ed.* 1996, 34, 2698–2700.

22. Zhao, W., Leroy, F., Heggen, B., Zahn, S., Kirchner, B., Balasubramanian, S., Muller- Plathe, F. Are there stable ion-pairs in room-temperature ionic liquids? Molecular dynamics simulations of 1-n-butyl-3-methylimidazolium hexafluorophosphate. *J. Am. Chem. Soc.* 2009, 131, 15825–15833.
23. Tsuzuki, S., Tokuda, H., Mikami, M. Theoretical analysis of the hydrogen bond of imidazolium C-2-H with anions. *Phys. Chem. Chem. Phys.* 2007, 9, 4780–4784.
24. Dupont, J., Scholten, J. D. On the structural and surface properties of transition-metal nanoparticles in ionic liquids. *Chem. Soc. Rev.* 2010, 39, 1780–1804.
25. Anderson, J. L., Armstrong, D. W. High-stability ionic liquids. A new class of stationary phases for gas chromatography. *Anal. Chem.* 2003, 75, 4851–4858.
26. Schroder, U., Wadhawan, J. D., Compton, R. G., Marken, F., Suarez, P. A. Z., Consorti, C. S., de Souza, R. F., Dupont, J. Water-induced accelerated ion diffusion: voltammetric studies in 1-methyl3-[2,6-(S)-dimethylocten-2-yl]imidazolium tetrafluoroborate, 1-butyl-3-methylimidazolium tetrafluoroborate and hexafluorophosphate ionic liquids. *New J. Chem.* 2000, 24, 1009–1015.
27. Chaumont, A., Schurhammer, R., Wipff, G. Aqueous interfaces with hydrophobic room-temperature ionic liquids: A molecular dynamics study. *J. Phys. Chem. B* 2005, 109, 18964–18973.
28. Chaumont, A., Wipff, G. Solvation of fluoro and mixed fluoro/chloro complexes of Eu-III in the [BMI][PF_6] room temperature ionic liquid. A theoretical study. *Phys. Chem. Chem. Phys.* 2005, 7, 1926–1932.
29. Sieffert, N., Wipff, G. Rhodium-catalyzed hydroformylation of 1-hexene in an ionic liquid: A molecular dynamics study of the hexene/[BMI][PF_6] interface. *J. Phys. Chem. B* 2007, 111, 4951–4962.
30. Sebesta, R., Kmentova, I., Toma, S. Catalysts with ionic tag and their use in ionic liquids. *Green Chem.* 2008, 10, 484–496.
31. Lee, S. G. Functionalized imidazolium salts for task-specific ionic liquids and their applications. *Chem. Commun.* 2006, 1049–1063.
32. Sawant, A. D., Raut, D. G., Darvatkar, N. B., Salunkhe, M. M. Recent developments of task-specific ionic liquids in organic synthesis. *Green Chem. Lett. Rev.* 2011, 4, 41–54.
33. Dyson, P. J., Laurenczy, G., Ohlin, C. A., Vallance, J., Welton, T. Determination of hydrogen concentration in ionic liquids and the effect (or lack of) on rates of hydrogenation. *Chem. Commun.* 2003, 2418–2419.
34. Hallett, J. P., Welton, T. Room-temperature ionic liquids: Solvents for synthesis and catalysis. 2. *Chem. Rev.* 2011, 111, 3508–3576.
35. Widegren, J. A., Finke, R. G. A review of the problem of distinguishing true homogeneous catalysis from soluble or other metal-particle heterogeneous catalysis under reducing conditions. *J. Mol. Catal. A Chem.* 2003, 198, 317–341.
36. Ott, L. S., Finke, R. G. Transition-metal nanocluster stabilization for catalysis: A critical review of ranking methods and putative stabilizers. *Coord. Chem. Rev.* 2007, 251, 1075–1100.
37. Scholten, J. D. From Soluble to Supported Iridium Metal Nanoparticles: Active and Recyclable Catalysts for Hydrogenation Reactions. *Curr. Org. Chem.* 2013, 17, 348–363.
38. Dupont, J., Fonseca, G. S., Umpierre, A. P., Fichtner, P. F. P., Teixeira, S. R. Transition-metal nanoparticles in imidazolium ionic liquids: Recyclable catalysts for biphasic hydrogenation reactions. *J. Am. Chem. Soc.* 2002, 124, 4228–4229.

39. Fonseca, G. S., Umpierre, A. P., Fichtner, P. F. P., Teixeira, S. R., Dupont, J. The use of imidazolium ionic liquids for the formation and stabilization of Ir^0 and Rh^0 nanoparticles: Efficient catalysts for the hydrogenation of arenes. *Chem. Eur. J.* 2003, 9, 3263–3269.
40. Huang, J., Jiang, T., Han, B. X., Gao, H. X., Chang, Y. H., Zhao, G. Y., Wu, W. Z. Hydrogenation of olefins using ligand-stabilized palladium nanoparticles in an ionic liquid. *Chem. Commun.* 2003, 1654–1655.
41. Rossi, L. M., Machado, G., Fichtner, P. F. P., Teixeira, S. R., Dupont, J. On the use of ruthenium dioxide in 1-n-butyl-3-methylimidazolium ionic liquids as catalyst precursor for hydrogenation reactions. *Catal. Lett.* 2004, 92, 149–155.
42. Umpierre, A. P., Machado, G., Fecher, G. H., Morais, J., Dupont, J. Selective hydrogenation of 1,3-butadiene to 1-butene by Pd(0) nanoparticles embedded in imidazolium ionic liquids. *Adv. Synth. Catal.* 2005, 347, 1404–1412.
43. Ott, L. S., Cline, M. L., Deetlefs, M., Seddon, K. R., Finke, R. G. Nanoclusters in ionic liquids: Evidence for N-heterocyclic carbene formation from imidazolium-based ionic liquids detected by ^2H NMR. *J. Am. Chem. Soc.* 2005, 127, 5758–5759.
44. Durand, J., Teuma, E., Malbosc, F., Kihn, Y., Gomez, M. Palladium nanoparticles immobilized in ionic liquid: An outstanding catalyst for the Suzuki C-C coupling. *Catal. Commun.* 2008, 9, 273–275.
45. Scheeren, C. W., Domingos, J. B., Machado, G., Dupont, J. Hydrogen reduction of Adams' catalyst in ionic liquids: Formation and stabilization of Pt(0) nanoparticles. *J. Phys. Chem. C* 2008, 112, 16463–16469.
46. Prechtl, M. H. G., Scariot, M., Scholten, J. D., Machado, G., Teixeira, S. R., Dupont, J. Nanoscale Ru(0) particles: Arene hydrogenation catalysts in imidazolium ionic liquids. *Inorg. Chem.* 2008, 47, 8995–9001.
47. Prechtl, M. H. G., Scholten, J. D., Dupont, J. Tuning the selectivity of ruthenium nanoscale catalysts with functionalised ionic liquids: Hydrogenation of nitriles. *J. Mol. Catal. A Chem.* 2009, 313, 74–78.
48. Marcilla, R., Mecerreyes, D., Odriozola, I., Pomposo, J. A., Rodriguez, J., Zalakain, I., Mondragon, I. New amine functional ionic liquid as building block in nanotechnology. *Nano* 2007, 2, 169–173.
49. Zhao, L., Zhang, C. Y., Zhuo, L., Zhang, Y. G., Ying, J. Y. Imidazolium salts: A mild reducing and antioxidative reagent. *J. Am. Chem. Soc.* 2008, 130, 12586–12587.
50. Zhang, H., Cui, H. Synthesis and characterization of functionalized ionic liquid-stabilized metal (gold and platinum) nanoparticles and metal nanoparticle/carbon nanotube hybrids. *Langmuir* 2009, 25, 2604–2612.
51. Yu, L., Sun, H., He, J., Wang, D., Jin, X., Hu, M., Chen, G. Z. Electro-reduction of cuprous chloride powder to copper nanoparticles in an ionic liquid. *Electrochem. Commun.* 2007, 9, 1374–1381.
52. Choi, S., Kim, K. S., Yeon, S. H., Cha, J. H., Lee, H., Kim, C. J., Yoo, I. D. Fabrication of silver nanoparticles via self-regulated reduction by 1-(2-hydroxyethyl)-3-methylimidazolium tetrafluoroborate. *Korean J. Chem. Eng.* 2007, 24, 856–859.
53. Redel, E., Thomann, R., Janiak, C. First correlation of nanoparticle size-dependent formation with the ionic liquid anion molecular volume. *Inorg. Chem.* 2008, 47, 14–16.

54. Torimoto, T., Okazaki, K., Kiyama, T., Hirahara, K., Tanaka, N., Kuwabata, S. Sputter deposition onto ionic liquids: Simple and clean synthesis of highly dispersed ultrafine metal nanoparticles. *Appl. Phys. Lett.* 2006, 89, 243117.
55. Okazaki, K. I., Kiyama, T., Hirahara, K., Tanaka, N., Kuwabata, S., Torimoto, T. Single-step synthesis of gold-silver alloy nanoparticles in ionic liquids by a sputter deposition technique. *Chem. Commun.* 2008, 691–693.
56. Khatri, O. P., Adachi, K., Murase, K., Okazaki, K., Torimoto, T., Tanaka, N., Kuwabata, S., Sugimura, H. Self-assembly of ionic liquid (BMI-PF_6)-stabilized gold nanoparticles on a silicon surface: Chemical and structural aspects. *Langmuir* 2008, 24, 7785–7792.
57. Suzuki, T., Okazaki, K., Kiyama, T., Kuwabata, S., Torimoto, T. A facile synthesis of AuAg alloy nanoparticles using a chemical reaction induced by sputter deposition of metal onto ionic liquids. *Electrochemistry* 2009, 77, 636–638.
58. Tsuda, T., Kurihara, T., Hoshino, Y., Kiyama, T., Okazaki, K.-I., Torimoto, T., Kuwabata, S. Electrocatalytic activity of platinum nanoparticles synthesized by room-temperature ionic liquid–sputtering method. *Electrochemistry* 2009, 77, 693–695.
59. Wender, H., de Oliveira, L. F., Migowski, P., Feil, A. F., Lissner, E., Prechtl, M. H. G., Teixeira, S. R., Dupont, J. Ionic liquid surface composition controls the size of gold nanoparticles prepared by sputtering deposition. *J. Phys. Chem. C* 2010, 114, 11764–11768.
60. Wender, H., Migowski, P., Feil, A. F., de Oliveira, L. F., Prechtl, M. H. G., Leal, R., Machado, G., Teixeira, S. R., Dupont, J. On the formation of anisotropic gold nanoparticles by sputtering onto a nitrile functionalised ionic liquid. *Phys. Chem. Chem. Phys.* 2011, 13, 13552–13557.
61. Gelesky, M. A., Umpierre, A. P., Machado, G., Correia, R. R. B., Magno, W. C., Morais, J., Ebeling, G., Dupont, J. Laser-induced fragmentation of transition metal nanoparticles in ionic liquids. *J. Am. Chem. Soc.* 2005, 127, 4588–4589.
62. Kimura, Y., Takata, H., Terazima, M., Ogawa, T., Isoda, S. Preparation of gold nanoparticles by the laser ablation in room-temperature ionic liquids. *Chem. Lett.* 2007, 36, 1130–1131.
63. Wender, H., Andreazza, M. L., Correia, R. R. B., Teixeira, S. R., Dupont, J. Synthesis of gold nanoparticles by laser ablation of an Au foil inside and outside ionic liquids. *Nanoscale* 2011, 3, 1240–1245.
64. Scholten, J. D., Leal, B. C., Dupont, J. Transition metal nanoparticle catalysis in ionic liquids. *ACS Catal.* 2012, 2, 184–200.
65. Gutel, T., Garcia- Anton, J., Pelzer, K., Philippot, K., Santini, C. C., Chauvin, Y., Chaudret, B., Basset, J. M. Influence of the self-organization of ionic liquids on the size of ruthenium nanoparticles: Effect of the temperature and stirring. *J. Mater. Chem.* 2007, 17, 3290–3292.
66. Migowski, P., Machado, G., Texeira, S. R., Alves, M. C. M., Morais, J., Traverse, A., Dupont, J. Synthesis and characterization of nickel nanoparticles dispersed in imidazolium ionic liquids. *Phys. Chem. Chem. Phys.* 2007, 9, 4814–4821.
67. Gutel, T., Santini, C. C., Philippot, K., Padua, A., Pelzer, K., Chaudret, B., Chauvin, Y., Basset, J. M. Organized 3D-alkyl imidazolium ionic liquids could be used to control the size of in situ generated ruthenium nanoparticles? *J. Mater. Chem.* 2009, 19, 3624–3631.

68. Salas, G., Podgorsek, A., Campbell, P. S., Santini, C. C., Padua, A. A. H., Gomes, M. F. C., Philippot, K., Chaudret, B., Turmine, M. Ruthenium nanoparticles in ionic liquids: Structural and stability effects of polar solutes. *Phys. Chem. Chem. Phys.* 2011, 13, 13527–13536.
69. Migowski, P., Zanchet, D., Machado, G., Gelesky, M. A., Teixeira, S. R., Dupont, J. Nanostructures in ionic liquids: Correlation of iridium nanoparticles' size and shape with imidazolium salts' structural organization and catalytic properties. *Phys. Chem. Chem. Phys.* 2010, 12, 6826–6833.
70. Silveira, E. T., Umpierre, A. P., Rossi, L. M., Machado, G., Morais, J., Soares, G. V., Baumvol, I. L. R., Teixeira, S. R., Fichtner, P. F. P., Dupont, J. The partial hydrogenation of benzene to cyclohexene by nanoscale ruthenium catalysts in imidazolium ionic liquids. *Chem. Eur. J.* 2004, 10, 3734–3740.
71. Redel, E., Kramer, J., Thomann, R., Janiak, C. Synthesis of Co, Rh and Ir nanoparticles from metal carbonyls in ionic liquids and their use as biphasic liquid-liquid hydrogenation nanocatalysts for cyclohexene. *J. Organomet. Chem.* 2009, 694, 1069–1075.
72. Vollmer, C., Redel, E., Abu- Shandi, K., Thomann, R., Manyar, H., Hardacre, C., Janiak, C. Microwave irradiation for the facile synthesis of transition-metal nanoparticles (NPs) in ionic liquids (ILs) from metal carbonyl precursors and Ru-, Rh-, and Ir-NP/IL dispersions as biphasic liquid-liquid hydrogenation nanocatalysts for cyclohexene. *Chem. Eur. J.* 2010, 16, 3849–3858.
73. Fonseca, G. S., Domingos, J. B., Nome, F., Dupont, J. On the kinetics of iridium nanoparticles formation in ionic liquids and olefin hydrogenation. *J. Mol. Catal. A Chem.* 2006, 248, 10–16.
74. Crabtree, R. Iridium compounds in catalysis. *Acc. Chem. Res.* 1979, 12, 331–338.
75. Prechtl, M. H. G., Campbell, P. S., Scholten, J. D., Fraser, G. B., Machado, G., Santini, C. C., Dupont, J., Chauvin, Y. Imidazolium ionic liquids as promoters and stabilising agents for the preparation of metal(0) nanoparticles by reduction and decomposition of organometallic complexes. *Nanoscale* 2010, 2, 2601–2606.
76. Dyson, P. J. Arene hydrogenation by homogeneous catalysts: Fact or fiction? *Dalton Trans.* 2003, 2964–2974.
77. Widegren, J. A., Finke, R. G. A review of soluble transition-metal nanoclusters as arene hydrogenation catalysts. *J. Mol. Catal. A Chem.* 2003, 191, 187–207.
78. Fonseca, G. S., Scholten, J. D., Dupont, J. Iridium nanoparticles prepared in ionic liquids: An efficient catalytic system for the hydrogenation of ketones. *Synlett* 2004, 1525–1528.
79. Campbell, P. S., Santini, C. C., Bayard, F., Chauvin, Y., Colliere, V., Podgorsek, A., Costa Gomes, M. F., Sa, J. Olefin hydrogenation by ruthenium nanoparticles in ionic liquid media: Does size matter? *J. Catal.* 2010, 275, 99–107.
80. Jutz, F., Andanson, J. M., Baiker, A. A green pathway for hydrogenations on ionic liquid-stabilized nanoparticles. *J. Catal.* 2009, 268, 356–366.
81. Cimpeanu, V., Kocevar, M., Parvulescu, V. I., Leitner, W. Preparation of rhodium nanoparticles in carbon dioxide induced ionic liquids and their application to selective hydrogenation. *Angew. Chem. Int. Ed.* 2009, 48, 1085–1088.
82. Venkatesan, R., Prechtl, M. H. G., Scholten, J. D., Pezzi, R. P., Machado, G., Dupont, J. Palladium nanoparticle catalysts in ionic liquids: Synthesis, characterisation and selective partial hydrogenation of alkynes to Z-alkenes. *J. Mater. Chem.* 2011, 21, 3030–3036.

83. Yuan, X., Yan, N., Xiao, C. X., Li, C. N., Fei, Z. F., Cai, Z. P., Kou, Y., Dyson, P. J. Highly selective hydrogenation of aromatic chloronitro compounds to aromatic chloroamines with ionic-liquid-like copolymer stabilized platinum nanocatalysts in ionic liquids. *Green Chem.* 2010, 12, 228–233.
84. Mu, X. D., Evans, D. G., Kou, Y. A. A general method for preparation of PVP-stabilized noble metal nanoparticles in room temperature ionic liquids. *Catal. Lett.* 2004, 97, 151–154.
85. Le Bras, J., Mukherjee, D. K., Gonzalez, S., Tristany, M., Ganchegui, B., Moreno-Manas, M., Pleixats, R., Henin, F., Muzart, J. Palladium nanoparticles obtained from palladium salts and tributylamine in molten tetrabutylammonium bromide: Their use for hydrogenolysis-free hydrogenation of olefins. *New J. Chem.* 2004, 28, 1550–1553.
86. Leger, B., Denicourt- Nowicki, A., Olivier- Bourbigou, H., Roucoux, A. Rhodium nanocatalysts stabilized by various bipyridine ligands in nonaqueous ionic liquids: Influence of the bipyridine coordination modes in arene catalytic hydrogenation. *Inorg. Chem.* 2008, 47, 9090–9096.
87. Dykeman, R. R., Yan, N., Scopelliti, R., Dyson, P. J. Enhanced rate of arene hydrogenation with imidazolium functionalized bipyridine stabilized rhodium nanoparticle catalysts. *Inorg. Chem.* 2011, 50, 717–719.
88. Leger, B., Denicourt- Nowicki, A., Olivier- Bourbigou, H., Roucoux, A. Rhodium colloidal suspensions stabilised by poly-N-donor ligands in non-aqueous ionic liquids: Preliminary investigation into the catalytic hydrogenation of arenes. *ChemSusChem* 2008, 1, 984–987.
89. Leger, B., Denicourt- Nowicki, A., Olivier- Bourbigou, H., Roucoux, A. Imidazolium-functionalized bipyridine derivatives: A promising family of ligands for catalytical Rh(0) colloids. *Tetrahedron Lett.* 2009, 50, 6531–6533.
90. Denicourt- Nowicki, A., Leger, B., Roucoux, A. N-Donor ligands based on bipyridine and ionic liquids: An efficient partnership to stabilize rhodium colloids. Focus on oxygen-containing compounds hydrogenation. *Phys. Chem. Chem. Phys.* 2011, 13, 13510–13517.
91. Xiao, C. X., Wang, H. Z., Mu, X. D., Kou, Y. Ionic-liquid-like copolymer stabilized nanocatalysts in ionic liquids - I. Platinum catalyzed selective hydrogenation of o-chloronitrobenzene. *J. Catal.* 2007, 250, 25–32.
92. Mu, X. D., Meng, J. Q., Li, Z. C., Kou, Y. Rhodium nanoparticles stabilized by ionic copolymers in ionic liquids: Long lifetime nanocluster catalysts for benzene hydrogenation. *J. Am. Chem. Soc.* 2005, 127, 9694–9695.
93. Zhao, C., Wang, H. Z., Yan, N., Xiao, C. X., Mu, X. D., Dyson, P. J., Kou, Y. Ionic-liquid-like copolymer stabilized nanocatalysts in ionic liquids: II. Rhodium-catalyzed hydrogenation of arenes. *J. Catal.* 2007, 250, 33–40.
94. Yang, X., Yan, N., Fei, Z. F., Crespo- Quesada, R. M., Laurenczy, G., Kiwi-Minsker, L., Kou, Y., Li, Y. D., Dyson, P. J. Biphasic hydrogenation over PVP stabilized Rh nanoparticles in hydroxyl functionalized ionic liquids. *Inorg. Chem.* 2008, 47, 7444–7446.
95. Hu, Y., Yu, Y. Y., Hou, Z. S., Li, H., Zhao, X. G., Feng, B. Biphasic hydrogenation of olefins by functionalized ionic liquid-stabilized palladium nanoparticles. *Adv. Synth. Catal.* 2008, 350, 2077–2085.
96. Kernchen, U., Etzold, B., Korth, W., Jess, A. Solid catalyst with ionic liquid layer (SCILL) - A new concept to improve selectivity illustrated by hydrogenation of cyclooctadiene. *Chem. Eng. Technol.* 2007, 30, 985–994.

97. Steinruck, H. P., Libuda, J., Wasserscheid, P., Cremer, T., Kolbeck, C., Laurin, M., Maier, F., Sobota, M., Schulz, P. S., Stark, M. Surface science and model catalysis with ionic liquid-modified materials. *Adv. Mater.* 2011, 23, 2571–2587.
98. Sobota, M., Happel, M., Amende, M., Paape, N., Wasserscheid, P., Laurin, M., Libuda, J. Ligand effects in SCILL model systems: Site-specific interactions with Pt and Pd nanoparticles. *Adv. Mater.* 2011, 23, 2617–2621.
99. Kume, Y., Qiao, K., Tomida, D., Yokoyama, C. Selective hydrogenation of cinnamaldehyde catalyzed by palladium nanoparticles immobilized on ionic liquids modified-silica gel. *Catal. Commun.* 2008, 9, 369–375.
100. Abu-Reziq, R., Wang, D., Post, M., Alper, H. Platinum nanoparticles supported on ionic liquid-modified magnetic nanoparticles: Selective hydrogenation catalysts. *Adv. Synth. Catal.* 2007, 349, 2145–2150.
101. Chun, Y. S., Shin, J. Y., Song, C. E., Lee, S. G. Palladium nanoparticles supported onto ionic carbon nanotubes as robust recyclable catalysts in an ionic liquid. *Chem. Commun.* 2008, 942–944.
102. Huang, J., Jiang, T., Gao, H. X., Han, B. X., Liu, Z. M., Wu, W. Z., Chang, Y. H., Zhao, G. Y. Pd nanoparticles immobilized on molecular sieves by ionic liquids: Heterogeneous catalysts for solvent-free hydrogenation. *Angew. Chem. Int. Ed.* 2004, 43, 1397–1399.
103. Gelesky, M. A., Scheeren, C. W., Foppa, L., Pavan, F. A., Dias, S. L. P., Dupont, J. Metal nanoparticle/ionic liquid/cellulose: New catalytically active membrane materials for hydrogenation reactions. *Biomacromolecules* 2009, 10, 1888–1893.
104. Bohm, V. P. W., Herrmann, W. A. Coordination chemistry and mechanisms of metal-catalyzed C-C coupling reactions; Part 12 nonaqueous ionic liquids: Superior reaction media for the catalytic Heck-vinylation of chloroarenes. *Chem. Eur. J.* 2000, 6, 1017–1025.
105. Gaikwad, D. S., Park, Y., Pore, D. M. A novel hydrophobic fluorous ionic liquid for ligand-free Mizoroki-Heck reaction. *Tetrahedron Lett.* 2012, 53, 3077–3081.
106. Li, L. Y., Wang, J. Y., Wu, T., Wang, R. H. Click ionic liquids: A family of promising tunable solvents and application in Suzuki-Miyaura cross-coupling. *Chem. Eur. J.* 2012, 18, 7842–7851.
107. Oliveira, F. F. D., dos Santos, M. R., Lalli, P. M., Schmidt, E. M., Bakuzis, P., Lapis, A. A. M., Monteiro, A. L., Eberlin, M. N., Neto, B. A. D. Charge-tagged acetate ligands as mass spectrometry probes for metal complexes investigations: Applications in Suzuki and Heck phosphine-free reactions. *J. Org. Chem.* 2011, 76, 10140–10147.
108. Muzart, J. Ionic liquids as solvents for catalyzed oxidations of organic compounds. *Adv. Synth. Catal.* 2006, 348, 275–295.
109. Khumraksa, B., Phakhodee, W., Pattarawarapan, M. Rapid oxidation of organic halides with N-methylmorpholine N-oxide in an ionic liquid under microwave irradiation. *Tetrahedron Lett.* 2013, 54, 1983–1986.
110. Lissner, E., de Souza, W. F., Ferrera, B., Dupont, J. Oxidative desulfurization of fuels with task-specific ionic liquids. *ChemSusChem* 2009, 2, 962–964.
111. Climent, M. J., Corma, A., Iborra, S. Homogeneous and heterogeneous catalysts for multicomponent reactions. *RSC Adv.* 2012, 2, 16–58.
112. Domling, A., Wang, W., Wang, K. Chemistry and biology of multicomponent reactions. *Chem. Rev.* 2012, 112, 3083–3135.

113. Panda, S. S., Khanna, P., Khanna, L. Biginelli reaction: A green perspective. *Curr. Org. Chem.* 2012, 16, 507–520.
114. Isambert, N., Duque, M. D. S., Plaquevent, J. C., Genisson, Y., Rodriguez, J., Constantieux, T. Multicomponent reactions and ionic liquids: A perfect synergy for eco-compatible heterocyclic synthesis. *Chem. Soc. Rev.* 2011, 40, 1347–1357.
115. Rajesh, S. M., Perumal, S., Menendez, J. C., Pandian, S., Murugesan, R. Facile ionic liquid-mediated, three-component sequential reactions for the green, regio- and diastereoselective synthesis of furocoumarins. *Tetrahedron* 2012, 68, 5631–5636.
116. Karthikeyan, P., Aswar, S. A., Muskawar, P. N., Bhagat, P. R., Kumar, S. S. Development and efficient 1-glycyl-3-methyl imidazolium chloride-copper(II) complex catalyzed highly enantioselective synthesis of 3, 4-dihydropyrimidin-2(1H)-ones. *J. Organomet. Chem.* 2013, 723, 154–162.
117. Ramos, L. M., Tobio, A., dos Santos, M. R., de Oliveira, H. C. B., Gomes, A. F., Gozzo, F. C., de Oliveira, A. L., Neto, B. A. D. Mechanistic studies on Lewis acid catalyzed Biginelli reactions in ionic liquids: Evidence for the reactive intermediates and the role of the reagents. *J. Org. Chem.* 2012, 77, 10184–10193.
118. Ramos, L. M., Guido, B. C., Nobrega, C. C., Corrêa, J. R., Silva, R. G., de Oliveira, H. C. B., Gomes, A. F., Gozzo, F. C., Neto, B. A. D. The Biginelli reaction with an imidazolium-tagged recyclable iron catalyst: Kinetics, mechanism, and antitumoral activity. *Chem. Eur. J.* 2013, 19, 4156–4168.
119. Zhang, Y. G., Chan, J. Y. G. Sustainable chemistry: Imidazolium salts in biomass conversion and CO_2 fixation. *Energy Environ. Sci.* 2010, 3, 408–417.
120. Zakrzewska, M. E., Bogel-Lukasik, E., Bogel-Lukasik, R. Ionic liquid-Mediated formation of 5-hydroxymethylfurfural - A promising biomass-derived building block. *Chem. Rev.* 2011, 111, 397–417.
121. Stahlberg, T., Fu, W. J., Woodley, J. M., Riisager, A. Synthesis of 5-(hydroxymethyl)furfural in ionic liquids: Paving the way to renewable chemicals. *ChemSusChem* 2011, 4, 451–458.
122. Tadesse, H., Luque, R. Advances on biomass pretreatment using ionic liquids: An overview. *Energy Environ. Sci.* 2011, 4, 3913–3929.
123. Keßler, M. T., Scholten, J. D., Prechtl, M. H. G. Metal catalysts immobilized in ionic liquids: A couple with opportunities for fine chemicals derived from biomass. In *New and Future Developments in Catalysis: Hybrid Materials, Composites, and Organocatalysts*, ed. S. Suib, 243–264. 1st ed. Elsevier, Amsterdam, 2013.
124. Rinaldi, R., Palkovits, R., Schuth, F. Depolymerization of cellulose using solid catalysts in ionic liquids. *Angew. Chem. Int. Ed.* 2008, 47, 8047–8050.
125. Qi, X. H., Watanabe, M., Aida, T. M., Smith, R. L. Efficient catalytic conversion of fructose into 5-hydroxymethylfurfural in ionic liquids at room temperature. *ChemSusChem* 2009, 2, 944–946.
126. Villandier, N., Corma, A. One pot catalytic conversion of cellulose into biodegradable surfactants. *Chem. Commun.* 2010, 46, 4408–4410.
127. Qi, X. H., Watanabe, M., Aida, T. M., Smith, R. L. Efficient one-pot production of 5-hydroxymethylfurfural from inulin in ionic liquids. *Green Chem.* 2010, 12, 1855–1860.

128. Chidambaram, M., Bell, A. T. A two-step approach for the catalytic conversion of glucose to 2,5-dimethylfuran in ionic liquids. *Green Chem.* 2010, 12, 1253–1262.
129. Rinaldi, R., Meine, N., vom Stein, J., Palkovits, R., Schuth, F. Which controls the depolymerization of cellulose in ionic liquids: The solid acid catalyst or cellulose? *ChemSusChem* 2010, 3, 266–276.
130. Dee, S. J., Bell, A. T. A study of the acid-catalyzed hydrolysis of cellulose dissolved in ionic liquids and the factors influencing the dehydration of glucose and the formation of humins. *ChemSusChem* 2011, 4, 1166–1173.
131. Bose, S., Armstrong, D. W., Petrich, J. W. Enzyme-catalyzed hydrolysis of cellulose in ionic liquids: A green approach toward the production of biofuels. *J. Phys. Chem. B* 2010, 114, 8221–8227.
132. Wang, Y., Radosevich, M., Hayes, D., Labbe, N. Compatible ionic liquid-cellulases system for hydrolysis of lignocellulosic biomass. *Biotechnol. Bioeng.* 2011, 108, 1042–1048.
133. Raut, D., Wankhede, K., Vaidya, V., Bhilare, S., Darwatkar, N., Deorukhkar, A., Trivedi, G., Salunkhe, M. Copper nanoparticles in ionic liquids: Recyclable and efficient catalytic system for 1,3-dipolar cycloaddition reaction. *Catal. Commun.* 2009, 10, 1240–1243.
134. Zhu, Y. H., Kong, Z. N., Stubbs, L. P., Lin, H., Shen, S. C., Anslyn, E. V., Maguire, J. A. Conversion of cellulose to hexitols catalyzed by ionic liquid-stabilized ruthenium nanoparticles and a reversible binding agent. *ChemSusChem* 2010, 3, 67–70.
135. Julis, J., Holscher, M., Leitner, W. Selective hydrogenation of biomass derived substrates using ionic liquid-stabilized ruthenium nanoparticles. *Green Chem.* 2010, 12, 1634–1639.
136. Zhao, H. B., Holladay, J. E., Brown, H., Zhang, Z. C. Metal chlorides in ionic liquid solvents convert sugars to 5-hydroxymethylfurfural. *Science* 2007, 316, 1597–1600.
137. Hu, S. Q., Zhang, Z. F., Song, J. L., Zhou, Y. X., Han, B. X. Efficient conversion of glucose into 5-hydroxymethylfurfural catalyzed by a common Lewis acid $SnCl_4$ in an ionic liquid. *Green Chem.* 2009, 11, 1746–1749.
138. Chan, J. Y. G., Zhang, Y. G. Selective conversion of fructose to 5-hydroxymethylfurfural catalyzed by tungsten salts at low temperatures. *ChemSusChem* 2009, 2, 731–734.
139. Zhang, Z., Wang, Q., Xie, H., Liu, W., Zhao, Z. Catalytic conversion of carbohydrates into 5-hydroxymethylfurfural by germanium(IV) chloride in ionic liquids. *ChemSusChem* 2011, 4, 131–138.
140. Stahlberg, T., Sorensen, M. G., Riisager, A. Direct conversion of glucose to 5-(hydroxymethyl)furfural in ionic liquids with lanthanide catalysts. *Green Chem.* 2010, 12, 321–325.
141. Tong, X. L., Li, M. R., Yan, N., Ma, Y., Dyson, P. J., Li, Y. D. Defunctionalization of fructose and sucrose: Iron-catalyzed production of 5-hydroxymethylfurfural from fructose and sucrose. *Catal. Today* 2011, 175, 524–527.
142. Binder, J. B., Raines, R. T. Simple chemical transformation of lignocellulosic biomass into furans for fuels and chemicals. *J. Am. Chem. Soc.* 2009, 131, 1979–1985.
143. Tao, F. R., Song, H. L., Chou, L. J. Hydrolysis of cellulose by using catalytic amounts of $FeCl_2$ in ionic liquids. *ChemSusChem* 2010, 3, 1298–1303.

144. Tao, F. R., Song, H. L., Yang, J., Chou, L. J. Catalytic hydrolysis of cellulose into furans in $MnCl_2$-ionic liquid system. *Carbohydr. Polym.* 2011, 85, 363–368.
145. Kim, B., Jeong, J., Lee, D., Kim, S., Yoon, H. J., Lee, Y. S., Cho, J. K. Direct transformation of cellulose into 5-hydroxymethyl-2-furfural using a combination of metal chlorides in imidazolium ionic liquid. *Green Chem.* 2011, 13, 1503–1506.
146. Li, C. Z., Zhang, Z. H., Zhao, Z. B. K. Direct conversion of glucose and cellulose to 5-hydroxymethylfurfural in ionic liquid under microwave irradiation. *Tetrahedron Lett.* 2009, 50, 5403–5405.
147. Zhang, Z. H., Zhao, Z. B. K. Microwave-assisted conversion of lignocellulosic biomass into furans in ionic liquid. *Bioresour. Technol.* 2010, 101, 1111–1114.
148. Wang, P., Yu, H. B., Zhan, S. H., Wang, S. Q. Catalytic hydrolysis of lignocellulosic biomass into 5-hydroxymethylfurfural in ionic liquid. *Bioresour. Technol.* 2011, 102, 4179–4183.
149. Pidko, E. A., Degirmenci, V., van Santen, R. A., Hensen, E. J. M. Glucose activation by transient Cr^{2+} dimers. *Angew. Chem. Int. Ed.* 2010, 49, 2530–2534.
150. Pidko, E. A., Degirmenci, V., van Santen, R. A., Hensen, E. J. M. Coordination properties of ionic liquid-mediated chromium(II) and copper(II) chlorides and their complexes with glucose. *Inorg. Chem.* 2010, 49, 10081–10091.
151. Cao, Q., Guo, X. C., Yao, S. X., Guan, J., Wang, X. Y., Mu, X. D., Zhang, D. K. Conversion of hexose into 5-hydroxymethylfurfural in imidazolium ionic liquids with and without a catalyst. *Carbohydr. Res.* 2011, 346, 956–959.
152. Amarasekara, A. S., Owereh, O. S. Hydrolysis and decomposition of cellulose in Bronsted acidic ionic liquids under mild conditions. *Ind. Eng. Chem. Res.* 2009, 48, 10152–10155.
153. Jia, S. Y., Cox, B. J., Guo, X. W., Zhang, Z. C., Ekerdt, J. G. Cleaving the beta-O-4 bonds of lignin model compounds in an acidic ionic liquid, 1-H-3-methylimidazolium chloride: An optional strategy for the degradation of lignin. *ChemSusChem* 2010, 3, 1078–1084.
154. Long, J. X., Guo, B., Li, X. H., Jiang, Y. B., Wang, F. R., Tsang, S. C., Wang, L. F., Yu, K. M. K. One step catalytic conversion of cellulose to sustainable chemicals utilizing cooperative ionic liquid pairs. *Green Chem.* 2011, 13, 2334–2338.
155. Cao, Q., Guo, X. C., Guan, J., Mu, X. D., Zhang, D. K. A process for efficient conversion of fructose into 5-hydroxymethylfurfural in ammonium salts. *Appl. Catal. A Gen.* 2011, 403, 98–103.

Index

W

X

Y

Z

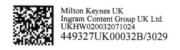

Milton Keynes UK
Ingram Content Group UK Ltd.
UKHW020032071024
449327UK00032B/3029

9 780367 262549